APPLIED RESEARCH IN FIELD CROP PATHOLOGY FOR INDIANA, 2022

APPLIED RESEARCH IN FIELD CROP PATHOLOGY FOR INDIANA, 2022

DARCY E. P. TELENKO
AND SUJOUNG SHIM

PURDUE UNIVERSITY PRESS
WEST LAFAYETTE, INDIANA

Cataloging-in-Publication Data on file at the Library of Congress.

978-1-62671-259-1 (paperback)

978-1-62671-260-7 (epdf)

CONTENTS

Acknowledgments   *ix*

Summary of 2022 Field Crop Disease Season   1

Agronomy Center for Research and Education (ACRE)   4

Corn   4

Evaluation of Fungicides for Foliar Disease in Corn in Central Indiana, 2022 (COR22-01.ACRE)   4

Evaluation of In-Furrow and Foliar Fungicides in Corn in Central Indiana, 2022 (COR22-15.ACRE)   6

Evaluation of Double Nickel in Corn in Central Indiana, 2022 (COR22-31.ACRE)   8

Evaluation of Fungicides for Foliar Disease in Corn in Central Indiana, 2022
(COR22-34.ACRE)   10

Comparison of Fungicide Efficacy of In-Furrow and 2x0 for Corn Diseases in Central Indiana, 2022
(COR22-37.ACRE)   12

Soybean   14

Comparison of Fungicide Efficacy for Foliar Disease of Soybeans in Central Indiana, 2022
(SOY22-01.ACRE)   14

Evaluation of Seed Treatment for Management of Sudden Death Syndrome on Soybean in Central
Indiana, 2022 (SOY22-03.ACRE)   16

Evaluation of Seed Treatment for Rhizoctonia in Soybean in Central Indiana, 2022
(SOY22-14.ACRE)   18

Uniform Oomycete Seed Treatment Trial in Soybean in Central Indiana, 2022
(SOY22-15.ACRE)   20

Evaluation of Fungicides for Soybean Foliar Diseases in Central Indiana, 2022
(SOY22-16.ACRE)   22

Evaluation of Foliar Fungicides in Soybean in Central Indiana, 2022 (SOY22-19.ACRE)   24

Evaluation of Foliar Fungicide Timing in Soybean in Central Indiana, 2022 (SOY22-22.ACRE)   26

Compare the Efficacy of In-Furrow Fungicide for Seedling Disease Soybean, 2022
(SOY22-24.ACRE)   28

Evaluation of Fungicides for Foliar Diseases in Soybean in Central Indiana, 2022
(SOY22-30.ACRE)   29

Evaluation of Fertilizers in Combination with Seed Treatments and Nano Products in Soybean in
Indiana, 2022 (SOY22-31.ACRE)   31

Wheat   33

Evaluation of Products and Cultivars for Fusarium Head Blight in Organic Wheat in Indiana, 2022
(WHT22-01.ACRE)   33

Evaluation of Foliar Fungicides for Scab management in Central Indiana, 2022
(WHT22-02.ACRE) 35

Evaluation of Foliar Fungicides and Cultivars for Scab Management in Central Indiana, 2022
(WHT22-03.ACRE) 37

Evaluation of Foliar Fungicides for Wheat Disease Management in Central Indiana, 2022
(WHT22-06.ACRE) 39

Evaluation of Foliar Fungicides for Wheat in Central Indiana, 2021 (WHT22-08_UFT.ACRE) 41

Pinney Purdue Agricultural Center (PPAC)   **43**

Corn 43

Uniform Fungicide Comparison for Tar Spot in Corn in Northwestern Indiana, 2022
(COR22-2_UFTTAR.PPAC) 43

Evaluation of Hybrids and Fungicide Timing for Tar Spot in Corn in Northwestern Indiana, 2022
(COR22-03.PPAC) 45

Evaluation of Products and Hybrids for Tar Spot in Organic Corn in Northwestern Indiana, 2022
(COR22-04.PPAC) 47

Fungicide Efficacy and Timing for Tar Spot in Corn in Northwestern Indiana, 2022
(COR22-05.PPAC) 49

Evaluation of Xyway Programs for Tar Spot Control in Northwestern Indiana, 2022
(COR22-14_UFTXYWAY.PPAC) 51

Fungicide Comparison for Tar Spot in Corn in Northwestern Indiana, 2022 (COR22-16.PPAC) 53

Evaluation of Fungicide Programs for Tar Spot in Corn in Northwestern Indiana, 2022
(COR22-18.PPAC) 55

Fungicide Comparison for Foliar Diseases in Corn in Northwestern Indiana, 2022
(COR22-24.PPAC) 57

Evaluation of Xyway Programs in Corn for Tar Spot in Northwestern Indiana, 2022
(COR22-27.PPAC) 59

Evaluation of Fungicide Timing and Application for Tar Spot in Corn in Northwestern Indiana, 2022
(COR22-29.PPAC) 61

Evaluation of Efficacy of CX-9032 and CX-10250 for Tar Spot in Corn in Northwestern Indiana,
2022 (COR22-30.PPAC) 63

Fungicide Timing and Application for Tar Spot in Corn in Northwestern Indiana, 2022
(COR22-32.PPAC) 65

Evaluation of Foliar Fungicides in Corn in Northwestern Indiana, 2022 (COR22-33.PPAC) 67

Evaluation of Drone Applications for Tar Spot in Corn in Northwestern Indiana, 2022
(COR22-35.PPAC) 69

Soybean 71

Fungicide Evaluation for White Mold in Soybean in Northwestern Indiana, 2022
(SOY22-04.PPAC) 71

Evaluation of Disease Management Options for White Mold in Organic Soybean in Northwestern
Indiana, 2022 (SOY22-06.PPAC) 73

Evaluation the Efficacy of Seed Treatments in Soybean in Northwestern Indiana, 2022 (SOY22-12.PPAC)   75

Evaluation of Fungicides for White Mold in Soybean in Northwestern Indiana, 2022 (SOY22-21.PPAC)   77

Evaluation of Fungicide Programs for White Mold in Soybean in Northwestern Indiana, 2022 (SOY22-23.PPAC)   79

Evaluation of Fungicides for White Mold in Soybean in Northwestern Indiana, 2022 (SOY22-26.PPAC)   81

Southwest Purdue Agricultural Center (SWPAC)   **83**

Soybean   83

Evaluation of Fungicides for Foliar Diseases on Soybean in Southwestern Indiana, 2022 (SOY22-02.SWPAC)   83

Evaluation of Fungicides for Foliar Diseases on Soybean in Southwestern Indiana, 2022 (SOY22-29.SWPAC)   85

Wheat   87

Evaluation of Foliar Fungicides for Scab Management in Southern Indiana, 2022 (WHT22-04.SWPAC)   87

Evaluation of Foliar Fungicides and Cultivars for Scab Management in Southern Indiana, 2022 (WHT22-05.SWPAC)   89

Davis Purdue Agricultural Center (DPAC)   **91**

Field-Scale Evaluation of Fungicides for Foliar Disease in Corn in Central Indiana, 2022 (COR22-08.DPAC)   91

Field-Scale Fungicide Timing Comparison for Foliar Diseases on Soybean in Central Indiana, 2022 (SOY22-07.DPAC)   93

Northeast Purdue Agricultural Center (NEPAC)   **95**

Field-Scale Fungicide Timing Comparison for Foliar Diseases on Corn in Northeastern Indiana, 2022 (COR22-09.NEPAC)   95

Evaluation of Xyway 2x2 Application for Foliar Diseases in Corn in Northeastern Indiana, 2022 (COR22-13.NEPAC)   97

Field-Scale Fungicide Timing for Foliar Diseases on Soybean in Northeastern Indiana, 2022 (SOY22-09.NEPAC)   99

Southeast Purdue Agricultural Center (SEPAC)   **101**

Field-scale Evaluation of Fungicide Timing for Foliar Disease in Corn in Southeastern Indiana, 2022 (COR22-10.SEPAC)   101

Field-scale Fungicide Timing Comparison for Foliar Diseases on Soybean in Southeastern Indiana, 2022 (SOY22-08.SEPAC)   103

Appendix: Weather Data   *105*

ACKNOWLEDGMENTS

This report is a summary of applied field crop pathology research trials conducted in 2022 under the direction of the Purdue Field Crop Pathology program in the Department of Botany and Plant Pathology at Purdue University. The authors wish to thank the Purdue Agronomy Research and Education Center, the Purdue Agricultural Centers, and the many cooperators and contributors who provided the resources needed to support the applied field crop pathology research program in Indiana. Special recognition is extended to Stephen Brand and Su Shim for technical skills in managing field trials, handling data organization and processing, and helping to prepare this report; Mariama Brown, Camila Da Silva, Monica Mizuno, and Kaitlin Waibel, graduate students and visiting scholars, who assisted with field trial data collection and analysis; Emily Duncan, Lindsey Berebitsky, and Ryan Gray, undergraduate student interns who assisted with field trial data collection and scouting; Dr. Tom Creswell, Dr. John Bonkowski, and Todd Abrahamson with the Purdue Plant Pest Diagnostic Laboratory for assistance in pathogen surveys and diagnosis. Collectively, the contributions of colleagues, professionals, students, and growers were responsible for a highly successful and productive program to evaluate products and practices for disease management in field crops.

The authors would also like to thank the following for their support in 2022: Adama, Bayer Crop Science, BASF, Biomineral Systems, LLC, Certis USA, Corteva Agriscience, FMC Agricultural Solution, Gowan, the Indiana Corn Marketing Council, the Indiana Soybean Alliance, the North Central Soybean Research Program, NC SARE Project # LNC20-443, Oro Agri, Pioneer, Purdue University, Sipcam Agro, Syngenta, the USDA NIFA Hatch Project #1019253, the USDA NIFA CARE Project #2021-09839, USWBSI-NFO, Valent, Vive Crop Protection, and VM Agritech.

SUMMARY OF 2022 FIELD CROP DISEASE SEASON

CORN

In 2022 there was low disease on corn in Indiana across the state; details of major issues are listed below. Gray leaf spot, northern corn leaf blight, northern corn leaf spot, and southern rust were found in pockets. There were also numerous reports of Physoderma brown spot and stalk rot. Tar spot and southern rust were two diseases that were closely monitored this season.

Tar spot: Tar spot of corn was a concern in 2022 due to previous epidemics. In 2022, very low levels of tar spot occurred in northern Indiana and in pockets in other areas of the state. The environmental conditions are key in determining field risk year to year, as leaf wetness plays an important role in tar spot disease development. The fourth year of tar spot–directed research has been completed here in Indiana. As a cautionary note, it is

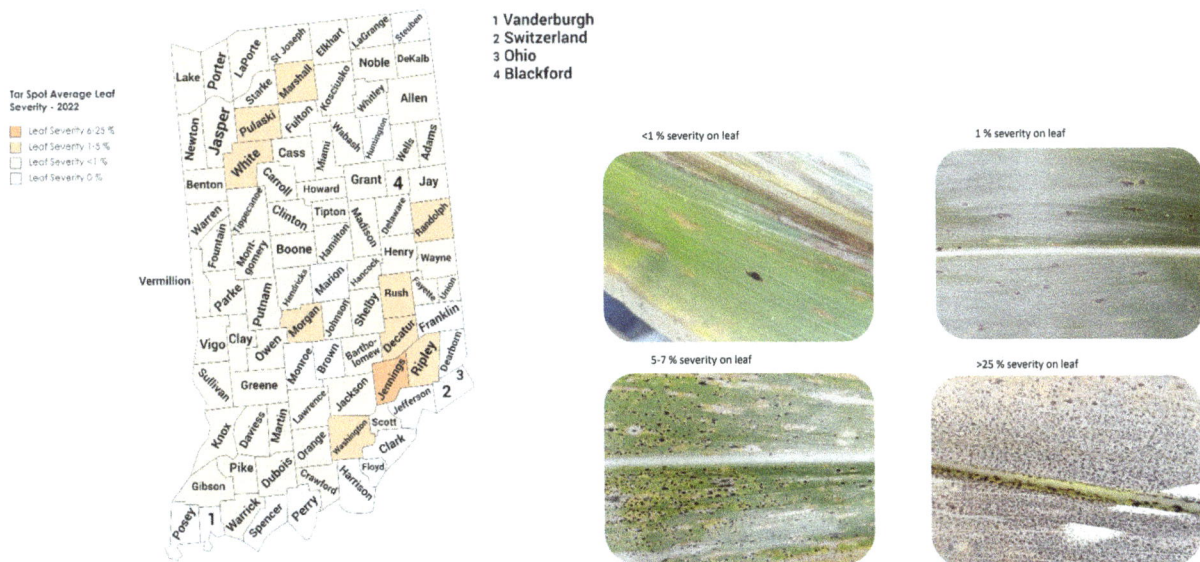

FIGURE 1. 2022 tar spot severity for Indiana. *Left:* The darker orange the county, the greater the severity of tar spot in the fields in which it was found. *Right (clockwise from top left):* The range of tar spot severity on leaves is >25%, 5–7%, 1% and <1%. Photo credit: D. Telenko. Map credit: D. Telenko.

FIGURE 2. *Left:* Southern corn rust map of confirmed (red) counties that had southern corn rust in Indiana in 2022 and a corn leaf with southern rust infection (*right*). Map source: https://corn.ipmpipe.org/southerncornrust/. Photo by D. Telenko.

still important to have multiple years of data for verification, but the initial results do serve as a good starting point for making future management decisions.

The field crop pathology team made a large effort at the end of the season to scout for tar spot across the state. Four new counties were confirmed with tar spot in 2022, making 86 counties total in Indiana to date. Out of the 201 fields scouted, 121 were positive for tar spot (60.2%). In addition, incidence and severity were rated (examples of severity are in Figure 1) and used to develop the severity map above—with increasing severity indicated by the darkness of the orange color of the county. The map demonstrates how tar spot development in 2022 was much lower than previous years. The map also parallels the weather conditions and reports during the season. It is important to document tar spot movement in the state so that when favorable conditions arise, increased tar spot disease risk can be more accurately assessed across the remainder of the state.

Southern corn rust. Southern corn rust was first found in Indiana in the 2022 season on August 12, and by the end of the season a total of 29 counties were confirmed to have the disease present (Figure 2, *left*). Southern rust pustules generally tend to occur on the upper surface of the leaf and produce chlorotic symptoms on the underside of the leaf (Figure 2, *right*). These pustules rupture the leaf surface and are orange to tan in color.

They are circular to oval in shape. Common rust was also widespread, and both diseases could be present on a leaf and easily mistaken for each other. It is important to send a sample to the Purdue Plant Pest Diagnostic Lab for confirmation if southern rust is suspected. There is an increased risk for yield impact if southern rust is identified early in the season.

Due to the need to monitor both southern rust and tar spot in Indiana, there will be **no charge for Indiana growers to submit southern rust and tar spot samples to the** Purdue Plant Pest Diagnostic Lab **for diagnostic confirmation.** This service is made possible through research supported by the Indiana Corn Marketing Council.

SOYBEAN

Diseases in soybeans remained relatively low throughout the season for much of the state. Our research sites and sentinel plots across the state saw low levels of frogeye leaf spot, Cercospora leaf blight, and Septoria brown spot. There were pockets where sudden death syndrome and white mold caused issues in fields. In general, it was a quiet year for foliar diseases in soybean.

WHEAT

Fusarium head blight (FHB), or scab, is one of the most impactful diseases of wheat and among the most challenging to prevent. In addition, FHB infection can cause the production of a mycotoxin called deoxynivalenol (DON), or vomitoxin. The conditions in 2022 were moderately conducive to FHB development. Our research sites in both West Lafayette and Vincennes had low levels of FHB develop in our nontreated susceptible cultivar checks, and initial DON testing was approximately 1 ppm. Fusarium head blight management requires an integrated approach, including selection of cultivars with moderate resistance and timely fungicide application at flowering. Very few other diseases were observed in our wheat trials.

AGRONOMY CENTER FOR RESEARCH AND EDUCATION (ACRE)

EVALUATION OF FUNGICIDES FOR FOLIAR DISEASE IN CORN IN CENTRAL INDIANA, 2022 (COR22-01.ACRE)

M. S. Mizuno, S. Shim, and D. E. P. Telenko, Department of Botany and Plant Pathology, Purdue University West Lafayette, IN 47907-2054

CORN (*ZEA MAYS* P0574AM)

Tar spot, *Phyllachora maydis*
Northern corn leaf blight, *Exserohilum turcicum*

A trial was established at the Purdue Agronomy Center for Research and Education (ACRE) in Tippecanoe County, Indiana. The trial was a randomized complete block design with six replications. Plots were 10 feet wide and 30 feet long and consisted of four rows, and the two center rows were used for evaluation. The previous crop was corn. Standard practices for nonirrigated grain corn production in Indiana were followed. Corn hybrid P0574AM was planted in 30-inch row spacing at a rate of 34,000 seeds/acre on May 31. All foliar fungicide applications were applied at 15 gal/acre and 40 psi using a Lee self-propelled sprayer equipped with a 10-foot boom, fitted with six TJ-VS 8002 nozzles spaced 20 inches apart. Fungicides were applied on July 22 at V10 and August 5 at tassel/silk (VT/R1) growth stages. Disease ratings were assessed on September 19 at dent (R5) growth stage. Tar spot stromata and northern corn leaf blight (NCLB) were rated by visually assessing the percentage (0–100%) of symptomatic leaf area on ear leaf of five plants in each plot. Values for the five leaves were averaged before analysis. Canopy green was rated by visually assessing the percentage (0–100%) of the whole plot for crop canopy that remained green at dent (R5) growth stage. The two center rows of each plot were harvested on October 19, and yields were adjusted to 15.5% moisture. All data were analyzed in SAS 9.4 (SAS Institute, Cary, NC). A generalized linear mixed model analysis of variance was performed using PROC GLIMMIX. Values are least squares means, and values with different letters are significantly different based on a least squares means test (α = 0.05).

In 2022, weather conditions were not favorable for disease development. Tar spot and NCLB were present but only remained at low levels. Veltyma and Trivapro applied at V10 and all treatments at VT/R1, except Headline AMP, significantly reduced NCLB severity over the nontreated controls (Table 1). Veltyma applied at VT/R1 significantly reduced tar spot stromata severity over the nontreated controls. No significant differences were observed for canopy greenness, harvest moisture, test weight, and yield of corn.

TABLE 1. *Effect of Treatment on Foliar Disease Severity, Canopy Greenness, and Yield of Corn*

TREATMENT, RATE/ACRE, AND TIMING[z]	NCLB[y] %	TAR SPOT[y] %	CANOPY GREEN[x] %	HARVEST MOISTURE %	TEST WEIGHT LB/BU	YIELD[w] BU/ACRE
Nontreated control 1	0.08 ab	0.14 abc	80.8	24.5	52.3	200.2
Headline AMP 1.68 SC 10.0 fl oz at V10	0.03 bcd	0.13 abc	82.5	24.8	52.8	208.0
Veltyma 3.34 S 7.0 fl oz at V10	0.01 cd	0.17 ab	87.5	24.9	52.3	201.7
Trivapro 2.21 SE 13.7 fl oz at V10	0.03 cd	0.12 a–d	87.5	24.8	52.5	206.0
Delaro Complete 458 SC 8.0 fl oz at V10	0.04 bc	0.11 a–d	81.7	24.4	54.2	201.6
Lucento 4.17 SC 5.0 fl oz at V10	0.04 bcd	0.19 a	85.8	23.0	53.0	206.4
Nontreated control 2	0.10 a	0.09 bcd	81.7	23.8	52.4	195.3
Headline AMP 1.68 SC 10.0 fl oz at VT/R1	0.04 bcd	0.09 b–e	87.5	24.4	52.6	210.4
Veltyma 3.34 S 7.0 fl oz at VT/R1	0.00 d	0.01 e	85.8	24.5	52.3	209.9
Trivapro 2.21 SE 13.7 fl oz at VT/R1	0.02 cd	0.10 bcd	85.8	25.1	52.5	207.3
Delaro Complete 458 SC 8.0 fl oz at VT/R1	0.03 cd	0.03 de	85.8	25.3	51.8	201.8
Lucento 4.17 SC 5.0 fl oz at VT/R1	0.01 cd	0.06 cde	87.5	25.1	52.3	208.4
P-value[v]	0.0006	0.0034	0.0636	0.2568	0.5795	0.1045

[z] Fungicide treatments were applied on July 22 at V10 and on August 5 at tassel/silk (VT/R1) growth stage.

[y] Foliar disease severity was visually assessed as a percentage (0–100%) of symptomatic leaf area on ear leaf, and five plants were assessed per plot and averaged before analysis on September 19. NCLB = northern corn leaf blight.

[x] Canopy greenness was visually assessed as a percentage (0–100%) of crop canopy green on September 19.

[w] Yields were adjusted to 15.5% moisture after harvest on October 19.

[v] All data were analyzed in SAS 9.4 (SAS Institute, Cary, NC). A generalized linear mixed model analysis of variance was performed using PROC GLIMMIX. Values are least squares means, and values with different letters are significantly different based on a least squares means test ($\alpha = 0.05$).

EVALUATION OF IN-FURROW AND FOLIAR FUNGICIDES IN CORN IN CENTRAL INDIANA, 2022 (COR22-15.ACRE)

I. L. Miranda, S. Shim, and D. E. P. Telenko, Department of Botany and Plant Pathology, Purdue University West Lafayette, IN 47907-2054

CORN (ZEA MAYS P0574AM)

Gray leaf spot, *Cercospora zeae-maydis*
Northern corn leaf blight, *Exserohilum turcicum*

A trial was established at the Purdue Agronomy Center for Research and Education (ACRE) in Tippecanoe County, Indiana. The experiment was a randomized complete block design with six replications. Plots were 10 feet wide and 30 feet long and consisted of four rows, and the two center rows were used for evaluation. The previous crop was corn. Standard practices for nonirrigated grain corn production in Indiana were followed. Corn hybrid P0574AM was planted in 30-inch row spacing at a rate of 2 seeds/foot on May 13. In-furrow applications were applied at planting at 10 gal/acre or 2xo application applied by CO_2 backpack sprayer at 10 gal/acre. Foliar applications were made at tassel/silk (VT/R1) growth stage on July 22. All foliar fungicide applications were applied at 15 gal/acre and 40 psi using a Lee self-propelled sprayer equipped with a 10-foot boom, fitted with six TJ-VS 8002 nozzles spaced 20 inches apart. Disease ratings were assessed on September 8 at dent (R5) growth stage. Gray leaf spot (GLS) and northern corn leaf blight (NCLB) severity were visually assessed as a percentage (0–100%) of symptomatic leaf area on ear leaf, and five plants were assessed per plot and averaged before analysis. The two center rows of each plot were harvested on October 15, and yields were adjusted to 15.5% moisture. All data were analyzed in SAS 9.4 (SAS Institute, Cary, NC). A generalized linear mixed model analysis of variance was performed using PROC GLIMMIX. Values are least squares means, and values with different letters are significantly different based on a least squares means test ($\alpha = 0.05$).

In 2022, weather conditions were moderately favorable for disease. GLS and NCLB were present in the trial but only remained at low levels. All treatments significantly reduced GLS over the nontreated control on September 8 except Xyway 2xo (Table 2). There was no significant effect of treatment on NCLB, harvest moisture, test weight, and yield of corn.

TABLE 2. *Effect of Treatment on Foliar Disease Severity and Yield of Corn*

TREATMENT, RATE/ACRE, AND TIMING[z]	GLS[y] %	NCLB[y] %	HARVEST MOISTURE %	TEST WEIGHT LB/BU	YIELD[x] BU/ACRE
Nontreated control	1.2 a	4.3	18.2	55.3	226.2
Xyway LFR 15.2 fl oz in-furrow	0.4 b	0.7	18.9	55.7	237.2
Xyway LFR 15.2 fl oz in 2xo	0.9 a	2.0	18.2	55.6	213.8
Xyway LFR 10.5 fl oz in-furrow fb Topguard EQ 5 fl oz at VT/R1	0.2 b	1.0	19.0	55.5	232.2
Topguard EQ 5.0 fl oz at VT/R1ç	0.1 b	0.0	18.8	55.3	276.2
Veltyma 3.34 S 7.0 fl oz at VT/R1	0.2 b	0.4	18.0	55.8	234.0
P-value[w]	0.0001	0.0844	0.2492	0.7369	0.7269

[z] In-furrow applications were applied at planting at 10 gal/acre or 2xo application applied by CO_2 backpack sprayer at 10 gal/acre. Foliar fungicide applications were made at tassel/silk (VT/R1) growth stage on July 22. fb = followed by.

[y] Foliar disease severity were visually assessed as a percentage (0–100%) of symptomatic leaf area on ear leaf, and five plants were assessed per plot and averaged before analysis on September 8. GLS = gray leaf spot, NCLB = northern corn leaf blight.

[x] Yields were adjusted to 15.5% moisture after harvest on October 15.

[w] All data were analyzed in SAS 9.4 (SAS Institute, Cary, NC). A generalized linear mixed model analysis of variance was performed using PROC GLIMMIX. Values are least squares means, and values with different letters are significantly different based on a least squares means test (α = 0.05).

EVALUATION OF DOUBLE NICKEL IN CORN IN CENTRAL INDIANA, 2022 (COR22-31.ACRE)

S. Shim and D. E. P. Telenko, Department of Botany and Plant Pathology, Purdue University
West Lafayette, IN 47907-2054

CORN (ZEA MAYS P0574AM)

Gray leaf spot, *Cercospora zeae-maydis*
Northern corn leaf blight, *Exserohilum turcicum*

A trial was established at the Purdue Agronomy Center for Research and Education (ACRE) in Tippecanoe County, Indiana. The experiment was a randomized complete block design with four replications. Plots were 10 feet wide and 30 feet long and consisted of four rows, and the two center rows were used for evaluation. The previous crop was corn. Standard practices for nonirrigated grain corn production in Indiana were followed. Corn hybrid P0574AM was planted in 30-inch row spacing at a rate of 34,000 seeds/feet on May 13. In-furrow applications were applied at planting at 10 gal/acre, and 2x0 applications were applied by CO_2 backpack sprayer on May 14. Stand counts were taken at V_4 growth stage on June 24. Disease ratings were assessed on September 8 at dent (R_5) growth stage. Northern corn leaf blight (NCLB) and gray leaf spot (GLS) were visually assessed as a percentage (0–100%) of symptomatic leaf area on ear leaf, and five plants were assessed per plot and averaged before analysis. The two center rows of each plot were harvested on October 15, and yields were adjusted to 15.5% moisture. All data were analyzed in SAS 9.4 (SAS Institute, Cary, NC). A generalized linear mixed model analysis of variance was performed using PROC GLIMMIX. Values are least squares means, and values with different letters are significantly different based on a least squares means test ($\alpha = 0.05$).

In 2022, weather conditions were not favorable for disease development. NCLB and GLS were present in the trial but only reached low levels. There was no significant effect of treatment on stand count compared to the nontreated control on June 24 (Table 3). There was no significant effect of treatment on GLS, NCLB, harvest moisture, test weight, and yield of corn (Table 3).

TABLE 3. *Effect of Treatment on Stand Count, Foliar Disease Severity, and Yield of Corn*

TREATMENT, RATE/ACRE, AND APPLICATION[z]	STAND COUNT #/A[y]	GLS[x] %	NCLB[x] %	HARVEST MOISTURE %	TEST WEIGHT LB/BU	YIELD[w] BU/ACRE
Nontreated control	36,881	1.1	8.0	17.9	56.0	195.5
Xyway LFR 10.5 fl oz in 2xo	35,678	0.4	4.4	17.8	56.4	191.9
Double Nickel LC 8.0 oz in-furrow	35,429	0.6	5.8	17.0	56.2	189.0
Double Nickel LC 8.0 oz in 2xo	31,508	0.8	3.7	18.4	56.3	195.1
Double Nickel LC 16.0 oz in-furrow	35,429	0.4	8.0	17.5	67.7	192.6
Double Nickel LC 16.0 oz in 2xo	35,356	0.6	6.4	17.5	56.5	194.0
P-value[v]	0.5439	0.0945	0.6338	0.7333	0.5860	0.8673

[z] In-furrow applications were applied at planting on May 13, and 2xo applications were applied on May 14 by CO_2 backpack sprayer.

[y] Stand counts were taken at V4 growth stage on June 24.

[x] Foliar disease severity was visually assessed as a percentage (0–100%) of symptomatic leaf area on ear leaf, and five plants were assessed per plot and averaged before analysis on September 8. NCLB = northern corn leaf blight, GLS = gray leaf spot.

[w] Yields were adjusted to 15.5% moisture after harvest on October 15.

[v] All data were analyzed in SAS 9.4 (SAS Institute, Cary, NC). A generalized linear mixed model analysis of variance was performed using PROC GLIMMIX. Values are least squares means, and values with different letters are significantly different based on a least squares means test (α = 0.05).

EVALUATION OF FUNGICIDES FOR FOLIAR DISEASES IN CORN IN CENTRAL INDIANA, 2022 (COR22-34.ACRE)

C. R. Da Silva, S. Shim, and D. E. P. Telenko, Department of Botany and Plant Pathology,
Purdue University West Lafayette, IN 47907-2054

CORN (*ZEA MAYS P0574AM*)

Tar spot, *Phyllachora maydis*
Gray leaf spot, *Cercospora zeae-maydis*
Northern corn leaf blight, *Exserohilum turcicum*

A trial was established at the Purdue Agronomy Center for Research and Education (ACRE) in Tippecanoe County, Indiana. The experiment was a randomized complete block design with four replications. Plots were 10 feet wide and 30 feet long and consisted of four rows, and the two center rows were used for evaluation. The previous crop was corn. Standard practices for grain corn production in Indiana were followed. Corn hybrid P0574AM was planted in 30-inch row spacing at a rate of 34,000 seeds/acre on May 13. All fungicide applications were applied at 15 gal/acre and 40 psi using a Lee self-propelled sprayer equipped with a 10-foot boom, fitted with six TJ-VS 8002 nozzles spaced 20 inches apart. Fungicides were applied on August 5 at silk (R1) growth stage. Disease ratings were assessed on September 15 at dent (R5) growth stage. Gray leaf spot (GLS), northern corn leaf blight (NCLB), and tar spot were rated by visually assessing the percent of severity on ear leaf on five plants in each plot. Values for the five leaves were averaged before analysis. The two center rows of each plot were harvested on October 19, and yields were adjusted to 15.5% moisture. All data were analyzed in SAS 9.4 (SAS Institute, Cary, NC). A generalized linear mixed model analysis of variance was performed using PROC GLIMMIX. Values are least squares means, and values with different letters are significantly different based on a least squares means test ($\alpha = 0.05$).

In 2022, weather conditions were not favorable for disease development. GLS and NCLB were the primary diseases, and tar spot was also present in the trial. There was no significant effect of treatment on tar spot over the nontreated controls (Table 4). All treatments reduced GLS and NCLB over the nontreated controls. No significant differences between treatments for harvest moisture, test weight, and yield of corn were observed.

TABLE 4. *Effect of Fungicide on Foliar Diseases Severity and Yield of Corn*

TREATMENT AND RATE/ACRE[z]	TAR SPOT[y] %	GLS[y] %	NCLB[y] %	HARVEST MOISTURE %	TEST WEIGHT LB/BU	YIELD[x] BU/ACRE
Nontreated control	0.00	0.7 a	2.7 a	23.7	52.5	197.9
ADM.03509.F.3.D 8.0 fl oz	0.00	0.1 b	1.0 b	23.6	52.3	206.8
ADM.03509.F.3.D 12.0 fl oz	0.00	0.1 b	0.7 b	23.3	52.7	214.0
ADM.03509.F.3.D 16.0 fl oz	0.03	0.1 b	1.0 b	23.8	52.3	204.4
ADM.03509.F.3.B 16.0 fl oz	0.03	0.1 b	1.0 b	23.5	52.6	206.6
Stratego YLD 4.65 fl oz	0.00	0.2 b	1.0 b	23.1	53.3	203.5
Headline AMP 1.68 SC 14.4 fl oz	0.00	0.2 b	0.5 b	23.1	52.3	202.8
Quilt Xcel 2.2 SE 14.0 fl oz	0.00	0.1 b	0.7 b	23.6	52.7	207.1
P-value[w]	0.5828	0.0001	0.0308	0.9646	0.6095	0.2875

[z] Fungicide treatments were applied on August 5 at silk (R1) growth stage.

[y] Foliar disease severity were visually assessed as a percentage (0–100%) of symptomatic leaf area on ear leaf, and five plants were assessed per plot and averaged before analysis on September 15. GLS = gray leaf spot, NCLB = northern corn leaf blight.

[x] Yields were adjusted to 15.5% moisture after harvest on October 19.

[w] All data were analyzed in SAS 9.4 (SAS Institute, Cary, NC). A generalized linear mixed model analysis of variance was performed using PROC GLIMMIX. Values are least squares means, and values with different letters are significantly different based on a least squares means test (α = 0.05).

COMPARISON OF FUNGICIDE EFFICACY OF IN-FURROW AND 2X0 FOR CORN DISEASES IN CENTRAL INDIANA, 2022 (COR22-37.ACRE)

C. R. Da Silva, S. Shim, and D. E. P. Telenko, Department of Botany and Plant Pathology, Purdue University West Lafayette, IN 47907-2054

CORN (*ZEA MAYS* P0574AM)

Gray leaf spot, *Cercospora zeae-maydis*
Northern corn leaf blight, *Exserohilum turcicum*

A trial was established at the Agronomy Center for Research and Education (ACRE) in Tippecanoe County, Indiana. The experiment was a randomized complete block design with four replications. Plots were 10 feet wide and 30 feet long and consisted of four rows, and the two center rows were used for evaluation. The previous crop was corn. Standard practices for nonirrigated grain corn production in Indiana were followed. Corn hybrid P0574AM was planted in 30-inch row spacing at a rate of 2 seeds/foot on May 13 with a plot planter. In-furrow applications were applied at plating at 10 gal/acre, and 2x0 applications were applied by CO_2 backpack sprayer on May 14 at 10 gal/acre. All foliar fungicide applications were applied at 15 gal/acre and 40 psi using a Lee self-propelled sprayer equipped with a 10-foot boom, fitted with six TJ-VS 8002 nozzles spaced 20 inches apart. Fungicides were applied on July 22 at tassel/silk (VT/R1) growth stage. Disease ratings were assessed on September 8 at dent (R5) growth stage. Gray leaf spot (GLS) and northern corn leaf blight (NCLB) was rated by visually assessing the percentage of disease per leaf on five plants in each plot at the ear leaf. Values for the five leaves were averaged before analysis. The two center rows of each plot were harvested on October 15, and yields were adjusted to 15.5% moisture. All data were analyzed in SAS 9.4 (SAS Institute, Cary, NC). A generalized linear mixed model analysis of variance was performed using PROC GLIMMIX. Values are least squares means, and values with different letters are significantly different based on a least squares means test ($\alpha = 0.05$).

In 2022, weather conditions were not favorable for disease development. GLS and NCLB were present in the trial but only remained at low levels. There was no significant effect on treatments on GLS and NCLB compared to Veltyma on September 8 (Table 5). There was no significant effect of treatment on harvest moisture, test weight, and yield of corn.

TABLE 5. *Effect of Treatment on Foliar Disease Severity and Yield of Corn*

TREATMENT, RATE/ACRE, AND TIMING[z]	GLS[y] %	NCLB[y] %	HARVEST MOISTURE %	TEST WEIGHT LB/BU	YIELD[x] BU/ACRE
Standard Veltyma 3.34 S 7.0 fl oz at VT/R1	0.1	0.05	19.6	55.2	211.8
Proline 480 SC 4.0 fl oz in-furrow fb Veltyma 3.34 S 7.0 fl oz at VT/R1	0.1	0.05	18.3	56.0	217.0
Xyway LFR 15.2 fl oz 2xo fb Veltyma 3.34 S 7.0 fl oz at VT/R1	0.1	0.00	19.2	55.4	223.1
Azteroid FC 3.3 4.2 fl oz in-furrow fb Veltyma 3.34 S 7.0 fl oz at VT/R1	0.1	0.10	18.9	55.4	206.4
Azteroid FC 3.3 8.4 fl oz in 2xo fb Veltyma 3.34 S 7.0 fl oz at VT/R1	0.1	0.03	19.2	54.6	210.5
P-value[w]	*0.8293*	*0.7088*	*0.5680*	*0.5061*	*0.1441*

[z] In-furrow applications were applied at plating on May 13 at 10 gal/acre, and 2xo applications were applied by CO_2 backpack sprayer on May 14. Foliar fungicide treatments were applied on July 22 at tassel/silk (VT/R1) growth stage. fb = followed by.

[y] Foliar disease severity were visually assessed as a percentage (0–100%) of symptomatic leaf area on ear leaf, and five plants were assessed per plot and averaged before analysis on September 8. GLS = gray leaf spot, NCLB = northern corn leaf blight.

[x] Yields were adjusted to 15.5% moisture after harvest on October 15.

[w] All data were analyzed in SAS 9.4 (SAS Institute, Cary, NC). A generalized linear mixed model analysis of variance was performed using PROC GLIMMIX. Values are least squares means, and values with different letters are significantly different based on a least squares means test (α = 0.05).

COMPARISON OF FUNGICIDE EFFICACY FOR FOLIAR DISEASE OF SOYBEANS IN CENTRAL INDIANA, 2022 (SOY22-01.ACRE)

E. A. Duncan, S. Shim, and D. E. P. Telenko, Department of Botany and Plant Pathology, Purdue University West Lafayette, IN 47907-2054

SOYBEAN (*GLYCINE MAX* P29A19E)

Septoria brown spot, *Septoria glycines*

A trial was conducted at the Purdue Agronomy Center for Research and Education (ACRE) in Tippecanoe County, Indiana. The experiment was a randomized complete block design with four replications. Plots were 10 feet wide and 30 feet long and consisted of four rows, and the two center rows were utilized for evaluation. The previous crop was corn. Standard practices for soybean production in Indiana were followed. Soybean cultivar P29A19E was planted in 30-inch row spacing at a rate of 140,000 seeds/acre on May 31. Fungicide applications were applied on August 5, at beginning pod/full pod (R3/R4) growth stage and were applied at 15 gal/acre at 40 psi using a CO_2 backpack sprayer equipped with a 10-foot boom, fitted with six TJ-VS 8002 nozzles spaced 20 inches apart. Disease ratings were assessed on September 14 at full seed (R6) growth stage. Septoria brown spot (SBS) was rated for disease severity by visually assessing the percentage of symptomatic leaf area in the lower canopy. Green stem was visually rated on a scale of 0–100% on October 4. The two center rows of each plot were harvested on October 7, and yields were adjusted to 13% moisture. All data were analyzed using a mixed model analysis of variance (SAS 9.4). Values are least squares means, and values with the same letter are not significantly different based on a least squares means test ($\alpha = 0.05$).

In 2022, weather conditions were not favorable for disease development. SBS was the most prominent disease in the trial and reached low severity. Only Miravis Neo reduced SBS over the nontreated control, but it was not significantly different from the other fungicides except Quadris (Table 6). Applications of Delaro Complete and Revytek resulted in increased green stem at harvest when compared to the nontreated control and other treatments. There were no differences between treatments for harvest moisture, test weight, and yield of soybean.

TABLE 6. *Effect of Treatment on Foliar Disease Severity, % Green Stem, and Yield of Soybean*

TREATMENT AND RATE/ACRE[z]	SBS[x] %	GREEN STEM[x] %	HARVEST MOISTURE %	TEST WEIGHT LB/BU	YIELD[w] BU/ACRE
Nontreated control	6.8 b	0.0 b	11.3	55.8	59.6
Topguard EQ 5.0 fl oz	4.0 bc	0.0 b	11.1	57.1	56.6
Lucento 4.17 SC 5.0 fl oz	4.5 bc	0.0 b	11.5	56.8	59.2
Trivapro 2.21 SE 13.7 fl oz	5.3 bc	0.3 b	12.0	56.7	52.0
Quadris 2.1 F 6.0 fl oz	13.5 a	0.0 b	11.4	56.8	55.8
Veltyma 3.34 S 7.0 fl oz	3.5 bc	0.0 b	11.9	56.8	51.8
Revytek 3.33 LC 8.0 fl oz	2.5 bc	1.3 a	12.3	56.6	55.7
Echo 36.0 fl oz + Folicur 3.6 F 4.0 fl oz +Topsin 20.0 fl oz	4.8 bc	0.0 b	12.3	56.6	55.4
Delaro Complete 3.83 SC 8.0 fl oz	3.5 bc	1.0 a	12.4	55.9	53.4
Miravis Neo 2.5 SE 13.7 fl oz	1.8 c	0.0 b	12.2	56.6	56.1
P-value[v]	0.0010	0.0044	0.4267	0.5767	0.5695

[z] Fungicide applications were made on August 5 at beginning pod/full pod (R3/R4) growth stage and contained a nonionic surfactant (Preference) at a rate of 0.25% v/v.

[y] Foliar disease incidence was rated on a scale of 0–100% of plants within a plot with disease symptoms. SBS = Septoria brown spot.

[x] Green stem was visually rated on a scale of 0–100% on October 4.

[w] Yields were adjusted to 13% moisture after harvest on October 7.

[v] All data were analyzed in SAS 9.4 (SAS Institute, Cary, NC). A generalized linear mixed model analysis of variance was performed using PROC GLIMMIX. Values are least squares means, and values with the same letter are not significantly different based on a least squares means test (α = 0.05).

EVALUATION OF SEED TREATMENT FOR MANAGEMENT OF SUDDEN DEATH SYNDROME ON SOYBEAN IN CENTRAL INDIANA, 2022 (SOY22-03.ACRE)

M. T. Brown, S. Shim, and D. E. P. Telenko, Department of Botany and Plant Pathology, Purdue University West Lafayette, IN 47907-2054

SOYBEAN (*GLYCINE MAX* P33A53X)

Sudden death syndrome, *Fusarium virguliforme*

A trial was established at the Purdue Agronomy Center for Research and Education (ACRE) in Tippecanoe County, Indiana. The experiment was a randomized complete block design with four replications. Plots were 10 feet wide and 30 feet long and consisted of four rows, and the two center rows were used for evaluation. The previous crop was corn. Standard practices for soybean production in Indiana were followed. Soybean cultivar P33A53X was planted in 30-inch row spacing at a rate of 140,000 seeds/acre on May 24. *Fusarium virguliforme* inoculum was applied at planting at 1.25 g/foot within the seedbed. Seed treatments were applied on seeds before planting. All treatments contained a base treatment except the nontreated control. Ten roots per plot were sampled from border rows at the full pod (R4) growth stage on August 12, gently washed, and rated for root rot severity on a scale of 0–100%. The two center rows of each plot were harvested on October 10, and yields were adjusted to 13% moisture. All data were analyzed using a mixed model analysis of variance. Values are least squares means, and values with different letters are significantly different based on a least squares means test ($\alpha = 0.05$).

In 2022, sudden death syndrome (SDS) root rot symptoms were evident, but weather conditions were not favorable for the development of SDS foliar symptoms (Table 7). There were no significant differences between the nontreated control and base and other seed treatments for root rot severity, green stem, and yield of soybean.

TABLE 7. *Effect of Seed Treatment on Sudden Death Syndrome (SDS) Root Rot Severity, Green Stem, and Yield of Soybean*

TREATMENT[z]	ROOT ROT[y] %	GREEN STEM[x] %	HARVEST MOISTURE %	TEST WEIGHT LB/BU	YIELD[w] BU/ACRE
Nontreated control	37.9	0.3	11.0	57.6	56.5
BASF Base	40.0	0.3	11.9	57.0	64.0
BASF Base + ILeVO (0.15 mg ai/seed)	38.9	0.5	11.6	57.0	63.2
BASF Base + Saltro (0.075 mg ai/seed)	40.1	0.8	11.3	57.2	59.4
BASF Base + Thiabendazole (0.64 fl oz/cwt) + Heads Up (0.16 fl oz/cwt) + BIO$_{ST}$ (0.16 fl oz/cwt) + Ascribe SAR (0.5 fl oz/cwt)	37.2	0.3	11.2	57.4	60.3
BASF Base + Saltro (0.075 mg ai/seed) + Ataplan (0.068 mg ai/seed)	40.1	0.3	11.2	57.3	62.9
BASF Base + CeraMax (2.46 fl oz/100 lbs)	35.0	0.5	11.4	57.3	55.8
BASF Base + ILeVO (0.15 mg ai/seed) + CeraMax (2.46 fl oz/100 lbs)	36.7	0.3	12.0	57.2	61.2
P-value[v]	0.7586	0.7709	0.4378	0.6554	0.4573

[z] Seed treatments were applied on seeds before planting. The BASF base contained Allegiance Fl at 4.0 g ai/100 kg, Stamina at 7.5 g ai/100 kg, Systiva XS Xemium Brand at 5.0 g ai/100 kg, and Poncho 600 at 0.11 mg ai/seed.

[y] Ten roots per plot were sampled from border rows at R4, gently washed, and rated for root rot severity on scale of 0–100% on August 12.

[x] Green stem was rated on a scale of 0–100% of stems remaining green within a plot on October 10.

[w] Yields were adjusted to 13% moisture after harvest on October 10.

[v] All data were analyzed in SAS 9.4 (SAS Institute, Cary, NC). A generalized linear mixed model analysis of variance was performed using PROC GLIMMIX. Values are least squares means, and values with different letters are significantly different based on a least squares means test (α = 0.05).

EVALUATION OF SEED TREATMENT FOR RHIZOCTONIA IN SOYBEAN IN CENTRAL INDIANA, 2022 (SOY22-14.ACRE)

S. Shim and D. E. P. Telenko, Department of Botany and Plant Pathology, Purdue University
West Lafayette, IN 47907-2054

SOYBEAN (*GLYCINE MAX* AG33XF2)

Rhizoctonia seedling blight, *Rhizoctonia solani*

A trial was established at the Purdue Agronomy Center for Research and Education (ACRE) in Tippecanoe County, Indiana. The experiment was a randomized complete block design with four replications. Plots were 10 feet wide and 30 feet long and consisted of four rows, and the two center rows were used for evaluation. The previous crop was corn. Standard practices for soybean production in Indiana were followed. Soybean cultivar AG33XF2 was planted in 30-inch row spacing at a rate of 140,000 seeds/acre on May 24. Seed treatments were applied by the cooperator. *Rhizocotonia solani* was inoculated in-furrow at planting at 1.25 g/foot. Stand counts were assessed on June 6 and June 13 at 14 days after planting and V1 growth stages, respectively. Canopy green was visually assessed as a percentage (0–100%) of canopy green on September 30. The two center rows of each plot were harvested on October 7, and yields were adjusted to 13% moisture. All data were analyzed in SAS 9.4 (SAS Institute, Cary, NC). A generalized linear mixed model analysis of variance was performed using PROC GLIMMIX. Values are least squares means, and values with different letters are significantly different based on a least squares means test (α = 0.05).

In 2022, weather conditions were not favorable for disease development. There was no significant effect of treatment on stand count, canopy greenness, harvest moisture, test weight, and yield of soybean (Table 8).

TABLE 8. *Effect of Treatment on Stand Counts, Canopy Greenness, and Yield of Soybean*

TREATMENT AND RATE/ACRE[z]	STAND COUNT #/A[y] JUNE 6	STAND COUNT #/A[y] JUNE 13	CANOPY GREEN[x] %	HARVEST MOISTURE %	TEST WEIGHT LB/BU	YIELD[w] BU/ACRE
Noninoculated, Zeltera Suite	97,357	94,307	12.5	11.7	56.2	63.4
Inoculated, Zeltera Suite System	91,912	91,258	12.5	11.6	56.3	62.5
Inoculated, Cruiser MAXX Vibrance	96,485	102,802	12.5	11.5	56.6	64.4
Inoculated, Acceleron System	100,624	102,148	11.3	11.5	56.2	67.2
Inoculated, Zeltera Suite System + Aveo	97,139	98,010	13.8	11.7	56.3	61.9
P-value[v]	0.1756	0.0766	0.9440	0.8793	0.8095	0.1705

[z] Seed treatments were applied by the cooperator. Plots were inoculated with *R. solani* in-furrow at planting at 1.25 g/foot.

[y] Stand counts were assessed on June 6 and June 13 at 14 days after planting and V1 growth stages, respectively.

[x] Canopy green was visually assessed as a percentage (0–100%) of canopy green on September 30.

[w] Yields were adjusted to 13% moisture after harvest on October 7.

[v] All data were analyzed in SAS 9.4 (SAS Institute, Cary, NC). A generalized linear mixed model analysis of variance was performed using PROC GLIMMIX. Values are least squares means, and values with different letters are significantly different based on a least squares means test (α = 0.05).

UNIFORM OOMYCETE SEED TREATMENT TRIAL IN SOYBEAN IN CENTRAL INDIANA, 2022 (SOY22-15.ACRE)

S. Shim and D. E. P. Telenko, Department of Botany and Plant Pathology, Purdue University West Lafayette, IN 47907-2054

SOYBEAN (*GLYCINE MAX* AG31XF2)

A trial was established at the Purdue Agronomy Center for Research and Education (ACRE) in Tippecanoe County, Indiana. The experiment was a randomized complete block design with four replications. Plots were 10 feet wide and 30 feet long and consisted of four rows, and the two center rows were used for evaluation. The previous crop was corn. Standard practices for soybean production in Indiana were followed. Soybean cultivar AG31XF2 was planted in 30-inch row spacing at a rate of 140,000 seeds/acre on May 24. Seed treatments were applied by the cooperator. Stand counts were assessed on July 6 and July 22 at V_3 and R_3 growth stages, respectively. Green stem was visually rated on a scale of 0–100% of stems remaining green within a plot on October 4. The two center rows of each plot were harvested on October 5, and yields were adjusted to 13% moisture. All data were analyzed in SAS 9.4 (SAS Institute, Cary, NC). A generalized linear mixed model analysis of variance was performed using PROC GLIMMIX. Values are least squares means, and values with different letters are significantly different based on a least squares means test (α = 0.05).

In 2022, weather conditions were not favorable for disease development. There was no significant effect of treatment on stand count, green stem, harvest moisture, test weight, and yield of soybean (Table 9).

TABLE 9. *Effect of Treatment on Stand Counts, Green Stem, and Yield of Soybean*

TREATMENT[z]	STAND COUNT #/A[y] JULY 6	STAND COUN #/A[y] JULY 22	GREEN STEM[x] %	HARVEST MOISTURE %	TEST WEIGHT LB/BU	YIELD[w] BU/ACRE
Base	88,645	109,553	1.3	9.8	56.7	54.6
Intego + Zeltera + Precinct	87,338	102,584	1.0	9.8	56.8	53.4
CruiserMaxx Vibrance	88,645	113,474	1.0	9.7	56.7	56.3
CruiserMaxx Vibrance +Vayantis	83,767	99,693	1.3	9.7	56.6	56.5
Obvius Plus + Poncho 600	88,209	108,247	1.3	9.8	56.7	55.6
Acceleron	83,635	110,642	1.0	9.9	56.4	49.9
P-value[v]	0.7231	0.2769	0.6398	0.9374	0.9698	0.2621

[z] Seed treatments were applied by the cooperator.

[y] Stand counts were assessed on July 6 and July 22 at V3 and R3 growth stages, respectively.

[x] Green stem was visually rated on a scale of 0–100% of stems remaining green within a plot on October 4.

[w] Yields were adjusted to 13% moisture after harvest on October 5.

[v] All data were analyzed in SAS 9.4 (SAS Institute, Cary, NC). A generalized linear mixed model analysis of variance was performed using PROC GLIMMIX. Values are least squares means, and values with different letters are significantly different based on a least squares means test (α = 0.05).

EVALUATION OF FUNGICIDES FOR SOYBEAN FOLIAR DISEASES IN CENTRAL INDIANA, 2022 (SOY22-16.ACRE)

S. Shim and D. E. P. Telenko, Department of Botany and Plant Pathology, Purdue University West Lafayette, IN 47907-2054

SOYBEAN (*GLYCINE MAX* P29A19E)

Frogeye leaf spot, *Cercospora sojina*
Septoria brown spot, *Septoria glycines*
Cercospora leaf blight, *Cercospora kikuchii*

A trial was established at the Purdue Agronomy Center for Research and Education (ACRE) in Tippecanoe County, Indiana. The experiment was a randomized complete block design with four replications. Plots were 10 feet wide and 30 feet long and consisted of four rows, and the two center rows were used for evaluation. The previous crop was corn. Standard practices for soybean production in Indiana were followed. Soybean cultivar P29A19E was planted in 30-inch row spacing at a rate of 140,000 seeds/acre on May 31. Fungicide applications were applied on August 5 at beginning pod/full pod (R3/R4) growth stage and were applied at 15 gal/acre at 40 psi using a CO_2 backpack sprayer equipped with a 10-foot boom, fitted with six TJ-VS 8002 nozzles spaced 20 inches apart. Disease ratings were assessed on September 14 at full seed (R6) growth stage. Frogeye leaf spot (FLS), Cercospora leaf blight (CLB), and Septoria brown spot (SBS) were rated for disease severity by visually assessing the percentage of symptomatic leaf area in the upper and lower canopies. The two center rows of each plot were harvested on October 4, and yields were adjusted to 13% moisture. All data were analyzed in SAS 9.4 (SAS Institute, Cary, NC). A generalized linear mixed model analysis of variance was performed using PROC GLIMMIX. Values are least squares means, and values with different letters are significantly different based on a least squares means test ($\alpha = 0.05$).

In 2022, weather conditions were not favorable for disease development. FLS, CLB, and SBS were present in the trial but only reached low levels. All fungicides reduced SBS over the nontreated control on September 14 (Table 10). There was no significant effect of treatment on FLS, CLB severity, harvest moisture, test weight, and yield of soybean.

TABLE 10. *Effect of Treatment on Foliar Disease Severity and Yield of Soybean*

TREATMENT AND RATE/ACRE[z]	FLS[Y] %	CLB[Y] %	SBS[Y] %	HARVEST MOISTURE %	TEST WEIGHT LB/BU	YIELD[x] BU/ACRE
Nontreated control	0.3	4.8	3.3 a	12.1	55.6	55.4
Miravis Neo 2.4 SE 13.7 fl oz	0.0	4.8	1.8 b	13.9	54.5	54.8
Miravis Top 1.67 SC 13.7 fl oz	0.0	2.3	1.5 b	14.5	54.8	57.2
Miravis Neo 2.4 SE 13.7 fl oz + Endigo ZCX 4.0 fl oz	0.0	3.0	1.8 b	14.5	54.6	54.7
Miravis Top 1.67 SC 13.7 fl oz + Endigo ZCX 4.0 fl oz	0.0	4.8	1.0 b	13.8	54.7	57.0
P-value[w]	0.0519	0.7657	0.0010	0.1650	0.1826	0.9341

[z] Fungicide applications were made on August 5 at R3/R4 growth stage.

[y] Foliar disease incidence was rated on a scale of 0–100% of plants within a plot with disease symptoms on September 14. FLS = frogeye leaf spot, SBS = Septoria brown spot, CLB= Cercospora leaf blight.

[x] Yields were adjusted to 13% moisture after harvest on October 4.

[w] All data were analyzed in SAS 9.4 (SAS Institute, Cary, NC). A generalized linear mixed model analysis of variance was performed using PROC GLIMMIX. Values are least squares means, and values with different letters are significantly different based on a least squares means test (α = 0.05).

EVALUATION OF FOLIAR FUNGICIDES IN SOYBEAN IN CENTRAL INDIANA, 2022 (SOY22-19.ACRE)

S. Shim and D. E. P. Telenko, Department of Botany and Plant Pathology, Purdue University
West Lafayette, IN 47907-2054

SOYBEAN (*GLYCINE MAX* P29A19E)

Frogeye leaf spot, *Cercospora sojina*
Septoria brown spot, *Septoria glycines*
Cercospora leaf blight, *Cercospora kikuchii*

A trial was established at the Purdue Agronomy Center for Research and Education (ACRE) in Tippecanoe County, Indiana. The experiment was a randomized complete block design with four replications. Plots were 10 feet wide and 30 feet long and consisted of four rows, and the two center rows were used for evaluation. The previous crop was corn. Standard practices for soybean production in Indiana were followed. Soybean cultivar P29A19E was planted in 30-inch row spacing at a rate of 8 seeds/foot on May 24. Xyway 2xo application was applied with a CO_2 backpack sprayer at 10 gal/acre at planting. Fungicide applications were applied at V5/R1 growth stage on July 11 using a Lee self-propelled sprayer and at beginning pod/full pod (R3/R4) growth stage on August 5 using a CO_2 backpack sprayer. All fungicides were applied at 15 gal/acre at 40 psi using a 10-foot boom, fitted with six TJ-VS 8002 nozzles spaced 20 inches apart. Disease ratings were assessed on September 14 at full seed (R6) growth stage. Frogeye leaf spot (FLS), Septoria brown spot (SBS), and Cercospora leaf bight (CLB) were rated for disease severity by visually assessing the percentage of symptomatic leaf area in the upper and lower canopies. The two center rows of each plot were harvested on October 4, and yields were adjusted to 13% moisture. All data were analyzed in SAS 9.4 (SAS Institute, Cary, NC). A generalized linear mixed model analysis of variance was performed using PROC GLIMMIX. Values are least squares means, and values with different letters are significantly different based on a least squares means test ($\alpha = 0.05$).

In 2022, weather conditions were not favorable for disease development. FLS, SBS, and CLB were present in the trial but only reached low levels. Delaro Complete and the program of Topguard applied at V5 followed by Lucento at R3 resulted in the lowest severity of SBS as compared to the nontreated control, but they were not significantly different from Miravis Top at 13.7 fl oz, Adastrio at 7.0 fl oz, or Topguard at 5.0 fl oz. There was no significant effect of treatment on FLS and CLB severity (Table 11). There was no significant effect of treatment on test weight (Table 11). Delaro complete increased yield over the nontreated control but was not significantly different from Miravis Top or Lucento.

TABLE 11. *Effect of Treatment on Foliar Disease Severity and Yield of Soybean*

TREATMENT, RATE/ACRE, AND TIMING[z]	FLS[y] %	SBS[y] %	CLB[y] %	HARVEST MOISTURE %	TEST WEIGHT LB/BU	YIELD[x] BU/ACRE
Nontreated control	1.2	4.0 a	6.3	11.2 d	57.9	55.5 bc
Topguard 4.29 EQ 5.0 fl oz at R3/R4	0.0	2.3 bc	5.8	12.4 a	57.3	55.8 bc
Lucento 4.17 SC 5.0 fl oz at R3/R4	0.0	3.3 ab	8.5	11.2 d	57.5	59.4 ab
Adastrio 4.0 SC 7.0 fl oz at R3/R4	0.0	2.3 bc	4.5	12.0 abc	57.3	55.5 bc
Adastrio 4.0 SC 8.0 fl oz at R3/R4	0.0	3.8 a	4.5	12.3 ab	57.1	52.1 c
Xyway 15.20 fl oz at plant 2xo	0.0	3.8 a	5.3	11.9 abc	59.6	53.7 c
Topguard 5.0 fl oz at V5 fb						
Lucento 4.17 SC 5.0 fl oz at R3/R4	0.0	3.8 c	4.3	11.5 cd	57.4	56.0 bc
Miravis Top 1.67 SC 13.7 fl oz at R3/R4	1.0	2.0 bc	6.3	11.6 bcd	57.1	58.6 ab
Delaro Complete 458 SC 8.0 fl oz at R3/R4	0.0	1.5 c	5.5	11.7 bcd	56.8	61.3 a
P-value[w]	0.2008	0.0022	0.0674	0.0057	0.4743	0.0102

[z] Xyway 2xo was applied to plants on May 24. Foliar fungicides were applied on July 11 at V5 and August 5 at R3-R4 (beginning to full pod) growth stages and contained a nonionic surfactant (Preference) at a rate of 0.25% v/v. fb = followed by.

[y] Foliar disease incidence was rated on a scale of 0–100% of plants within a plot with disease symptoms on September 14. FLS = frogeye leaf spot, SBS = Septoria brown spot, CLB = Cercospora leaf blight.

[x] Yields were adjusted to 13% moisture after harvest on October 4.

[w] All data were analyzed in SAS 9.4 (SAS Institute, Cary, NC). A generalized linear mixed model analysis of variance was performed using PROC GLIMMIX. Values are least squares means, and values with different letters are significantly different based on a least squares means test (α = 0.05).

EVALUATION OF FOLIAR FUNGICIDE TIMING IN SOYBEAN IN CENTRAL INDIANA, 2022 (SOY22-22.ACRE)

S. Shim and D. E. P. Telenko, Department of Botany and Plant Pathology, Purdue University West Lafayette, IN 47907-2054

SOYBEAN (*GLYCINE MAX* P29A19E)

Frogeye leaf spot, *Cercospora sojina*

Septoria brown spot, *Septoria glycines*

Cercospora leaf blight, *Cercospora kikuchii* and other *spp.*

A trial was established at the Purdue Agronomy Center for Research and Education (ACRE) in Tippecanoe County, Indiana. The experiment was a randomized complete block design with four replications. Plots were 10 feet wide and 30 feet long and consisted of four rows, and the two center rows were used for evaluation. The previous crop was corn. Standard practices for soybean production in Indiana were followed. Soybean cultivar P29A19E was planted in 30-inch row spacing at a rate of 140,000 seed/acre on May 31. Foliar fungicide applications were applied on August 5 and August 18 at beginning pod/full pod (R3/R4) and beginning seed (R5) growth stages, respectively. Applications were made using a CO_2 backpack sprayer at 15 gal/acre at 40 psi using a 10-foot boom, fitted with six TJ-VS 8002 nozzles spaced 20 inches apart. Disease ratings were assessed on September 14 at full seed (R6) growth stage. Frogeye leaf spot (FLS), Septoria brown spot (SBS), and Cercospora leaf bight (CLB) were rated for disease severity by visually assessing the percentage of symptomatic leaf area in the upper and lower canopies. The two center rows of each plot were harvested on October 7, and yields were adjusted to 13% moisture. All data were analyzed in SAS 9.4 (SAS Institute, Cary, NC). A generalized linear mixed model analysis of variance was performed using PROC GLIMMIX. Values are least squares means, and values with different letters are significantly different based on a least squares means test (α = 0.05).

In 2022, weather conditions were not favorable for disease development. FLS, SBS, and CLB were present in the trial but only reached low levels. There was no significant difference between treatments and nontreated controls for FLS and CLB on September 14 (Table 12). All fungicide treatments significantly reduced SBS severity over the nontreated control 1 but not in nontreated control 2. All treatments reduced defoliation over the nontreated control 1 except Lucento at R5 (Table 12). There was no significant effect of treatment on harvest moisture, test weight, and yield of soybean.

TABLE 12. *Effect of Treatment on Foliar Disease Severity, Defoliation, and Yield of Soybean*

TREATMENT, RATE/ACRE, AND TIMING[z]	FLS[y] %	SBS[y] %	CLÇB[y] %	DEFOLIATION[x] %	HARVEST MOISTURE %	TEST WEIGHT LB/BU	YIELD[w] BU/ACRE
Nontreated control 1	0.0	13.0 a	8.3	15.0 a	11.3	56.9	50.8
Delaro Complete 458 SC 8.0 fl oz at R3/R4	0.0	2.8 b	6.3	6.5 c	11.7	57.0	55.5
Lucento 4.17 SC 5.0 fl oz at R3/R4	0.0	2.8 b	8.5	8.5 bc	11.1	56.9	55.3
Trivapro 2.21 SE 13.7 fl oz at R3/R4	0.0	4.5 b	1.8	3.3 c	11.7	56.6	59.7
Miravis Neo 2.5 SE 13.7 fl oz at R3/R4	0.0	2.3 b	4.5	6.3 c	11.7	56.7	59.9
Revytek 3.33 LC 8.0 fl oz at R3/R4	0.0	2.5 b	5.0	6.3 c	12.3	56.3	56.6
Delaro Complete 458 SC 8.0 fl oz at R5	0.0	3.5 b	4.0	6.5 c	11.5	57.1	59.9
Lucento 4.17 SC 5.0 fl oz at R5	0.0	4.3 b	8.5	12.5 ab	11.1	57.2	53.7
Trivapro 2.21 SE 13.7 fl oz at R5	0.0	6.0 b	6.8	7.5 bc	11.5	57.1	53.1
Miravis Neo 2.5 SE 13.7 fl oz at R5	0.0	2.5 b	5.0	7.3 bc	11.5	56.8	55.9
Revytek 3.33 LC 8.0 fl oz at R5	0.1	4.3 b	7.5	8.8 bc	11.7	56.8	55.1
Nontreated control 2	0.0	5.5 b	3.8	6.5 c	11.0	57.2	63.2
P-value[v]	0.4710	0.0002	0.1444	0.0260	0.1496	0.2122	0.2569

[z] Foliar fungicide applications were applied on August 5 and August 18 at beginning pod/full pod (R3/R4) and beginning seed (R5) growth stages, respectively. All treatments contain NIS 0.25% v/v.

[y] Foliar disease incidence was rated on a scale of 0–100% of plants within a plot with disease symptoms on September 14. FLS = frogeye leaf spot, SBS = Septoria brown spot., CLB= Cercospora leaf blight.

[x] Defoliation was rated on a scale of 0–100% within in a plot September 14.

[w] Yields were adjusted to 13% moisture after harvest on October 7.

[v] All data were analyzed in SAS 9.4 (SAS Institute, Cary, NC). A generalized linear mixed model analysis of variance was performed using PROC GLIMMIX. Values are least squares means, and values with different letters are significantly different based on a least squares means test (α = 0.05).

COMPARE THE EFFICACY OF IN-FURROW FUNGICIDE FOR SEEDLING DISEASE SOYBEAN, 2022 (SOY22-24.ACRE)

C. R. Da Silva, S. Shim, and D. E. P. Telenko, Department of Botany and Plant Pathology, Purdue University West Lafayette, IN 47907-2054

SOYBEAN (*GLYCINE MAX* P29A19E)

A trial was established at the Agronomy Center for Research and Education (ACRE) in Tippecanoe County, Indiana. The experiment was a randomized complete block design with four replications. Plots were 10 feet wide and 30 feet long and consisted of four rows, and the two center rows were used for evaluation. The previous crop was corn. Standard practices for soybean production in Indiana were followed. Soybean cultivar P29A19E was planted in 30-inch row spacing at a rate of 140,000 seeds/acre on May 24. Ethos and Double Nickel applications were applied in-furrow at 10 gal/acre at planting on May 24. Stand counts were assessed on June 9 and June 13 at 10 days and 14 days after emergence, respectively. Green stem was visually rated on a scale of 0–100% on October 4. The center rows of each plot were harvested on October 5, and yields were adjusted to 13% moisture. All data were analyzed in SAS 9.4 (SAS Institute, Cary, NC). A generalized linear mixed model analysis of variance was performed using PROC GLIMMIX. Values are least squares means, and values with different letters are significantly different based on a least squares means test (α = 0.05).

In 2022, weather conditions were not favorable for disease development. No significant differences between treatments and the nontreated control were detected for stand counts, green stem, harvest moisture, test weight, and yield of soybean (Table 13).

TABLE 13. *Effect of Treatment on Stand Counts, Green Stem, and Yield of Soybean*

TREATMENT AND RATE/ACRE[z,ç]	STAND COUNT #A[y] JUNE 9	STAND COUNT #A[y] JUNE 13	GREEN STEM[x] %	HARVEST MOISTURE %	TEST WEIGHT LB/BU	YIELD[w] BU/ACRE
Nontreated control	77,537	79,933	1.3	10.5	56.3	40.4
Ethos XB 4.0 fl oz in-furrow	77,755	81,022	1.0	9.9	56.6	41.9
Double Nickel LC 8.0 oz in-furrow	71,656	77,319	1.0	10.3	57.2	43.0
Double Nickel LC 16.0 oz in-furrow	77,755	85,378	1.0	10.2	57.0	42.4
P-value[v]	0.2924	0.4466	0.4363	0.3129	0.4939	0.9723

[z] Ethos and Double Nickel applications were applied in-furrow at 10 gal/acre at planting on May 24.

[y] Stand counts were assessed on June 9 and June 13 at 10 days and 14 days after emergence, respectively.

[x] Green stem was visually rated on a scale of 0–100% on October 4.

[w] Yields were adjusted to 13% moisture after harvest on October 5.

[v] All data were analyzed in SAS 9.4 (SAS Institute, Cary, NC). A generalized linear mixed model analysis of variance was performed using PROC GLIMMIX. Values are least squares means, and values with different letters are significantly different based on a least squares means test (α = 0.05).

EVALUATION OF FUNGICIDES FOR FOLIAR DISEASES IN SOYBEAN IN CENTRAL INDIANA, 2022 (SOY22-30.ACRE)

C. R. Da Silva, S. Shim, and D. E. P. Telenko, Department of Botany and Plant Pathology, Purdue University West Lafayette, IN 47907-2054

SOYBEAN (*GLYCINE MAX* P29A19E)

Frogeye leaf spot, *Cercospora sojina*
Septoria brown spot, *Septoria glycines*
Cercospora leaf blight, *Cercospora kikuchii*

A trial was established at the Agronomy Center for Research and Education (ACRE) in Tippecanoe County, Indiana. The experiment was a randomized complete block design with four replications. Plots were 10 feet wide and 30 feet long and consisted of four rows, and the two center rows were used for evaluation. The previous crop was corn. Standard practices for soybean production in Indiana were followed. Soybean cultivar P29A19E was planted in 30-inch row spacing at a rate of 140,000 seeds/acre on May 31. Fungicide applications were applied at 15 gal/acre and 40 psi using a Lee self-propelled sprayer equipped with a 10-foot boom, fitted with six TJ-VS 8002 nozzles spaced 20 inches apart. Fungicides were applied on August 5 at full pod (R4) growth stage. Disease ratings were assessed on September 14 at full seed (R6) growth stage. Frogeye leaf spot (FLS), Cercospora leaf blight (CLB), and Septoria brown (SBS) were rated for disease severity by visually assessing the percentage of symptomatic leaf area in the upper and lower canopies. Canopy greenness was visually assessed as a percentage (0–100%) of crop canopy green on September 14. The center rows of each plot were harvested on October 5, and yields were adjusted to 13% moisture. All data were analyzed in SAS 9.4 (SAS Institute, Cary, NC). A generalized linear mixed model analysis of variance was performed using PROC GLIMMIX. Values are least squares means, and values with different letters are significantly different based on a least squares means test (α = 0.05).

In 2022, weather conditions were not favorable for disease development. FLS, CLB, and SBS were present in the trial but only remained at low levels. There was no significant effect of treatments on FLS and CLB severity. However, all treatments reduced SBS over the nontreated control on September 14 (Table 14). Fungicides ADM.03509.F.3.B at 12.0 fl oz and 16.0 fl oz significantly increased canopy greenness over the nontreated control (Table 14). All treatments showed higher harvest moisture compared to the nontreated control. There was no significant effect of treatment on test weight and yield of soybean.

TABLE 14. *Effect of Treatment on Foliar Disease Severity, Canopy Greenness, and Yield of Soybean*

TREATMENT AND RATE/ACRE[z]	FLS[y] %	CLB[y] %	SBS[y] %	CANOPY GREEN[x] %	HARVEST MOISTURE %	TEST WEIGHT LB/BU	YIELD[w] BU/ACRE
Nontreated control	0.3	6.5	7.0 a	76.3 b	10.6 c	56.2	52.0
ADM.03509.F.3.B 8.0 fl oz	0.0	6.5	3.3 b	76.3 b	11.9 ab	55.6	50.5
ADM.03509.F.3.B 12.0 fl oz	0.0	3.8	2.5 bc	85.0 a	11.6 abc	56.0	59.1
ADM.03509.F.3.B 16.0 fl oz	0.0	6.5	2.3 bc	80.0 a	12.4 a	55.5	50.9
ADM.03509.F.3.D 16.0 fl oz	0.0	6.8	2.3 bc	78.8 b	12.2 ab	56.0	53.4
Stratego YLD 4.65 fl oz	0.0	7.0	3.5 b	76.3 b	11.3 abc	56.4	56.7
Miravis Neo 2.5 SE 13.7 fl oz	0.3	3.8	1.3 c	80.0 ab	12.3 ab	54.8	53.4
Quadris Top 1.67 SC 14.0 fl oz	0.8	4.8	4.0 b	80.0 ab	11.2 bc	55.9	53.0
P-value[v]	*0.5227*	*0.4102*	*0.0001*	*0.0294*	*0.0378*	*0.2233*	*0.1449*

[z] Fungicides were applied on August 5 at R4 (full pod) growth stage.

[y] Foliar disease severity was rated by visually assessing the percentage of symptomatic leaf area in the upper and lower canopies on September 14. FLS = frogeye leaf spot, CLB = Cercospora leaf blight, SBS = Septoria brown spot.

[x] Canopy greenness was visually assessed as a percentage (0–100%) of crop canopy green on September 14.

[w] Yields were adjusted to 13% moisture after harvest on October 5.

[v] All data were analyzed in SAS 9.4 (SAS Institute, Cary, NC). A generalized linear mixed model analysis of variance was performed using PROC GLIMMIX. Values are least squares means, and values with different letters are significantly different based on a least squares means test (α = 0.05).

EVALUATION OF FERTILIZERS IN COMBINATION WITH SEED TREATMENTS AND NANO PRODUCTS IN SOYBEAN IN INDIANA, 2022 (SOY22-31.ACRE)

S. Shim and D. E. P. Telenko, Department of Botany and Plant Pathology, Purdue University
West Lafayette, IN 47907-2054

SOYBEAN (GLYCINE MAX AG23XF2)

Sudden death syndrome, *Fusarium virguliforme*

A trial was established at the Purdue Agronomy Center for Research and Education (ACRE) in Tippecanoe County, Indiana. The experiment was a randomized complete block design with four replications. Plots were 10 feet wide and 30 feet long and consisted of four rows, and the two center rows were used for evaluation. The previous crop was corn. Standard practices for soybean production in Indiana were followed. Soybean cultivar AG23XF2 was planted in 30-inch row spacing at a rate of 140,000 seeds/acre on May 24. *Fusarium virguliforme* inoculum was applied at planting at 1.25 g/foot within the seedbed. An application of 28% nitrogen (N) at 51 gal (150 lbs N) was made prior to planting soybean. Seed treatments were applied on seeds before planting and contained base seed treatment. Foliar applications were applied at beginning bloom (R1) growth stage on July 12. Applications were applied at 15 gal/acre and 40 psi using a Lee self-propelled sprayer equipped with a 10-foot boom, fitted with six TJ-VS 8002 nozzles spaced 20 inches apart. Stand counts were assessed at V3 on June 20. Phytoxicity was visually rated on a scale of 0–100%. Ten roots per plot were sampled from border rows at R4 on August 11, gently washed, and rated for root rot severity on scale of 0–100%. The two center rows of each plot were harvested on October 4, and yields were adjusted to 13% moisture. All data were analyzed using a mixed model analysis of variance. Values are least squares means, and values with different letters are significantly different based on a least squares means test (α = 0.05).

In 2022, weather conditions were not favorable for disease development. There was no significant effect of treatment on stand count, phytoxicity, root rot, and yield of soybean (Table 15).

TABLE 15. *Effect of Treatment on Stand Count, Phytoxicity, Root Rot, and Yield of Soybean*

TREATMENT, RATE/ACRE, AND TIMING[z]	STAND COUNT #/A	PHYTO[y] %	ROOT ROT[x] %	HARVEST MOISTURE %	TEST WEIGHT LB/BU	YIELD[w] BU/ACRE
Base	102,148	6.3	12.4	10.1	56.6	56.8
Base fb NanoStress 4.0 fl oz + NanoN 4.0 fl oz at R1	104,980	6.3	11.5	10.1	56.5	58.5
Base + 150 lbs N prior to planting	100,188	5.0	10.7	10.1	56.4	62.1
Base + Ilevo	108,464	6.3	13.3	10.0	56.6	59.0
Base + Ilevo fb NanoStress 4.0 fl oz + NanoN 4.0 fl oz at R1	101,930	5.0	15.2	10.0	56.2	59.0
Base + Ilevo + 150 lbs N prior to planting	107,593	7.5	14.1	10.0	56.4	62.4
P-value[v]	0.7275	0.5915	0.5425	0.8492	0.6761	0.4035

[z] Seed treatments were preapplied to the seeds of the cultivar. *Fusarium virguliforme* inoculum was applied at planting at 1.25 g/foot within the seedbed. Seed treatments were applied on seeds before planting. Foliar applications were applied a beginning bloom (R1) growth stage on July 12. All treatments contained a base treatment. fb = followed by.

[y] Phytoxicity (Phyto) was visually rated on a scale of 0–100% across the plots on June 20.

[x] Ten roots per plot were sampled from border rows at R4 on August 11 and rated for root rot severity on scale of 0–100%.

[w] Yields were adjusted to 13% moisture after harvest on October 4.

[v] All data were analyzed in SAS 9.4 (SAS Institute, Cary, NC). A generalized linear mixed model analysis of variance was performed using PROC GLIMMIX. Values are least squares means, and values with different letters are significantly different based on a least squares means test (α = 0.05).

EVALUATION OF PRODUCTS AND CULTIVARS FOR FUSARIUM HEAD BLIGHT IN ORGANIC WHEAT IN INDIANA, 2022 (WHT22-01.ACRE)

C. R. Da Silva, S. Shim, and D. E. P. Telenko, Department of Botany and Plant Pathology, Purdue University West Lafayette, IN 47907-2054

WHEAT (*TRITICUM AESTIVUM* KASKASKIA AND HARPOON)

Fusarium head blight, *Fusarium graminearum*

A trial was established at the Purdue Agronomy Center for Research and Education (ACRE) in Tippecanoe County, Indiana. The experiment was a randomized complete block design with four replications. Plots were 7.5 feet wide and 20 feet long and consisted of 12 rows spaced 7.5 inches apart, and the center of each plot was used for evaluation. The previous crop was corn. Organic wheat cultivars Kaskaskia and Harpoon were planted in 7.5-inch row spacing using a drill on November 8, 2021. All fungicide applications were applied at 15 gal/acre and 40 psi using a CO_2 backpack sprayer equipped with a 10-foot boom, fitted with six TJ-VS 8002 nozzles spaced 20 inches apart and directed forward and backward at a 45-degree angle. Fungicides were applied on May 24 at Feekes growth stage 10.5.1. All plots were inoculated with a mixture of isolates of *Fusarium graminearum* endemic to Indiana on May 25, with a spore suspension (50,000 spores/ml) applied at 300 ml/plot. Disease ratings were assessed on June 13. Fusarium head blight (FHB) incidence was measured as the number of infected heads out of 60 plants in each plot and calculated as a percentage. FHB severity was rated by visually assessing the percentage (0–100%) of the infected heads. The FHB index was calculated as (% FHB incidence multiplied by average FHB severity)/100 per plot. The eight center rows of each plot were harvested with a Kincaid plot combine on July 7, and yields were adjusted to 13.5% moisture. A subsample of grain was taken from each plot and partitioned for deoxynivalenol (DON) analysis completed by the University of Minnesota DON testing lab and to determine Fusarium damage kernels (FDK) by visually assessing the percentage (0–100%) of the infected heads. Data were subjected to mixed model analysis of variance (SAS 9.4, 2019), and means were compared based on a least squares means test ($\alpha = 0.05$).

In 2022, weather conditions were moderately favorable for FHB. FHB was the most prominent disease. There were no significant interactions between cultivar and fungicide treatments; therefore, main effects of each are presented (Table 16). No differences were detected for FHB incidence and index in both the Harpoon and Kaskaskia cultivars. In the cultivar Harpoon, FHB severity was reduced when compared to Kaskaskia. There were no differences in foliar treatments from the nontreated control for FHB incidence, severity, and index. The concentration of DON was significantly lower in the cultivar Harpoon as compared to Kaskaskia and when treated with Prosaro. There was no significant difference between treatments and cultivars for Fusarium damaged kernels (FDK). The cultivar Harpoon had a highest percent of wheat yield when compared to Kaskaskia.

TABLE 16. *Effect of Cultivar and Fungicide on Fusarium Head Blight (FHB), Deoxynivalenol (DON), Fusarium Damaged Kernels (FDK), and Yield of Wheat*

TREATMENT[z]	FHB DI[y] %	FHB DS[x] %	FHB INDEX[w]	DON PPM[v]	FDK[u] %	YIELD[t] BU/ACRE
Cultivar						
Kaskaskia	13.0	7.9 a	0.8	0.377 a	10.1	44.4 b
Harpoon	17.9	4.1 b	0.7	0.215 b	9.2	51.6 a
Fungicide rate/A						
Nontreated control	13.8	9.4	1.2	0.296 a	10.0	47.9
Prosaro 421 SC 8.2 fl oz	13.3	3.6	0.4	0.110 b	9.5	49.1
ChampION 50 WP 1.5 lb	21.7	3.7	0.7	0.351 a	8.2	48.8
Pacesetter WS 13.0 fl oz	17.1	5.2	0.7	0.301 a	10.1	46.3
Sonata 1.0 qt	10.8	7.9	0.6	0.396 a	9.9	46.4
Actinovate AG 12.0 fl oz	16.0	6.3	0.7	0.308 a	10.3	49.5
P-value cultivar[s]	*0.0816*	*0.0150*	*0.5295*	*0.0007*	*0.1238*	*0.0001*
P-value fungicide	*0.3105*	*0.1587*	*0.3415*	*0.0285*	*0.3291*	*0.7328*
P-value cultivar*fungicide	*0.3438*	*0.2749*	*0.3321*	*0.2773*	*0.3879*	*0.0157*

[z] Fungicides were applied on May 24 at Feekes growth stage 10.5.1. All plots were inoculated with a mixture of isolates of *Fusarium graminearum* endemic to Indiana on May 25, with a spore suspension (50,000 spores/ml) applied at 300 ml/plot.

[y] FHB disease incidence (DI) was measured as the number of infected heads out of 60 plants in each plot and calculated as a percentage on June 13.

[x] FHB disease severity (DS) was rated by visually by assessing the percentage of the infected head.

[w] The FHB index was calculated as (% FHB incidence multiplied by average FHB severity)/100 per plot. FHB = Fusarium head blight.

[v] Analysis of mycotoxin deoxynivalenol (DON) was completed by the University of Minnesota DON Testing Lab.

[u] Fusarium damaged kernels (FDK) were visually assessed as a percentage (0–100%) of the infected heads.

[t] Yields were adjusted to 13.5% moisture after harvest on July 7.

[s] All data were analyzed in SAS 9.4 (SAS Institute, Cary, NC). A generalized linear mixed model analysis of variance was performed using PROC GLIMMIX. Values are least squares means, and values with different letters are significantly different based on a least squares means test (α = 0.05).

EVALUATION OF FOLIAR FUNGICIDES FOR SCAB MANAGEMENT IN CENTRAL INDIANA, 2022 (WHT22-02.ACRE)

K. M. Goodnight, S. Shim, and D. E. P. Telenko, Department of Botany and Plant Pathology, Purdue University West Lafayette, IN 47907-2054

WHEAT (*TRITICUM AESTIVUM* P25R40)

Fusarium head blight, *Fusarium graminearum*

A trial was established at the Purdue Agronomy Center for Research and Education (ACRE) in Tippecanoe County, Indiana. The experiment was a randomized complete block design with four replications. Plots were 7.5 feet wide and 20 feet long and consisted of 12 rows spaced 7.5 inches apart, and the center of each plot was used for evaluation. The previous crop was corn. On November 8, 2021, the wheat cultivar P25R40 was drilled at 7.5 inches spacing. All fungicide applications were applied at 15 gal/acre and 40 psi using a CO_2 backpack sprayer equipped with a 10-foot boom, fitted with six TJ-VS 8002 nozzles spaced 20 inches apart and directed forward and backward at a 45-degree angle. Fungicides were applied on May 24 and May 30 at Feekes growth stages 10.5.1 and 10.5.1 + 6 days after 10.5.1, respectively. All plots were inoculated with a mixture of isolates of *Fusarium graminearum* endemic to Indiana on May 25. The spore suspension (50,000 spores/ml) was applied at 300 ml/plot with the CO_2 handheld sprayer. Disease ratings were assessed on June 13. Fusarium head blight (FHB) incidence was measured as the number of infected heads out of 60 plants in each plot and calculated as a percentage. FHB severity was rated by visually assessing the percentage of the infected head. The FHB index was calculated as (% FHB incidence multiplied by average FHB severity)/100 per plot. Values for each plot were averaged before analysis. The eight center rows of each plot were harvested with a Kincaid plot combine on July 7, and yields were adjusted to 13.5% moisture. Data were subjected to mixed model analysis of variance (SAS 9.4, 2019), and means were compared using a least squares means test ($\alpha = 0.05$).

In 2022, weather conditions were moderately favorable for FHB. FHB was the most prominent disease. No differences were detected for FHB incidence and severity as compared to the nontreated control (Table 17). The FHB index was reduced by Miravis Ace applied at 10.5.1 and Miravis Ave at 10.5.1 followed by Sphaerex at 6 DAT over the nontreated control. The concentration of deoxynivalenol (DON) was reduced over the non-treated control by all treatments except for Prosaro applied at 10.5.1 (Table 17). Harvest moisture was higher in all of the fungicide-treated plots except for Prosaro Pro at 10.5.1, as compared to the nontreated control. There was no significant difference in yield of wheat.

TABLE 17. *Effect of Fungicide on Fusarium Head Blight (FHB), Deoxynivalenol (DON), Fusarium Damaged Kernels (FDK), and Yield of Wheat*

TREATMENT, RATE/ACRE, AND TIMING[z]	FHB DI[y] %	FHB DS[x] %	FHB INDEX[w]	DON[v] PPM	FDK[u] %	HARVEST MOISTURE %	YIELD[t] BU/ACRE
Nontreated control	20.4	4.7	0.9 ab	0.853 a	9.5	18.3 cd	57.3
Prosaro 421 SC 6.5 fl oz at 10.5.1	20.0	3.1	0.6 abc	0.585 ab	9.5	18.8 bcd	58.9
Caramba 90 EC 13.5 fl oz at 10.5.1	21.7	2.7	0.6 abc	0.525 b	9.5	19.2 ab	59.0
Miravis Ace 5.2 SC 13.7 fl oz at 10.5.1	10.4	2.6	0.3 c	0.280 bc	8.0	19.0 ab	57.8
Prosaro Pro 400 SC 10.3 fl oz at 10.5.1	25.0	3.8	1.0 a	0.513 b	7.8	18.3 d	55.6
Sphaerex 7.3 fl oz at 10.5.1	16.3	2.0	0.5 bc	0.340 bc	8.8	18.9 abc	56.6
Miravis Ace 5.2 SC 13.7 fl oz at 10.5.1 fb Prosaro Pro 400 SC 10.3 fl oz at 6 DAT	19.2	2.7	0.5 bc	0.345 bc	7.8	19.5 a	63.6
Miravis Ace 5.2 SC 13.7 fl oz at 10.5.1 fb Sphaerex 7.3 fl oz at 6 DAT	8.3	3.5	0.3 c	0.188 c	7.5	19.3 ab	54.5
Miravis Ace 5.2 SC 13.7 fl oz at 10.5.1 fb Tebuconazole 3.6 F 4.0 fl oz at 6 DAT	14.6	3.2	0.5 bc	0.355 bc	8.8	19.0 ab	56.3
P-value[s]	*0.0964*	*0.4660*	*0.0445*	*0.0081*	*0.1394*	*0.0060*	*0.0542*

[z] Fungicide treatments were applied on May 24 and May 30 at Feekes growth stages 10.5.1 and 10.5.1 + 6 days after treatment (DAT), respectively. All treatments contained a nonionic surfactant (Preference) at a rate of 0.25% v/v. All plots were inoculated with *Fusarium graminearum* spore suspension (50,000 spores/ml) after the treatment at Feekes 10.5.1. Spore suspension was applied at 300 ml/plot with a handheld sprayer on May 25. fb = followed by.

[y] FHB disease incidence (DI) was measured as the number of infected heads out of 60 plants in each plot and calculated as a percentage on June 13.

[x] FHB disease severity (DS) was rated by visually assessing the percentage of the infected head on June 13. FHB = Fusarium head blight.

[w] The FHB index was calculated as (total FHB incidence multiplied by average FHB severity)/100 per plot on June 13.

[v] Analysis of mycotoxin deoxynivalenol (DON) was completed by the University of Minnesota DON Testing Lab.

[u] Fusarium damaged kernels (FDK) were visually assessed as a percentage (0–100%) of the infected heads.

[t] Yields were adjusted to 13.5% moisture after harvest on July 7.

[s] All data were analyzed in SAS 9.4 (SAS Institute, Cary, NC). A generalized linear mixed model analysis of variance was performed using PROC GLIMMIX. Values are least squares means, and values with different letters are significantly different based on a least squares means test ($\alpha = 0.05$).

EVALUATION OF FOLIAR FUNGICIDES AND CULTIVARS FOR SCAB MANAGEMENT IN CENTRAL INDIANA, 2022 (WHT22-03.ACRE)

M. S. Mizuno, S. Shim, and D. E. P. Telenko, Department of Botany and Plant Pathology, Purdue University West Lafayette, IN 47907-2054

WHEAT (*TRITICUM AESTIVUM* P25R40 AND P25R61)

Fusarium head blight, *Fusarium graminearum*

A trial was established at the Purdue Agronomy Center for Research and Education (ACRE) in Tippecanoe County, Indiana. The experiment was a randomized complete block design with four replications. Plots were 7.5 feet wide and 20 feet long and consisted of 12 rows spaced 7.5 inches apart, and the center of each plot was used for evaluation. The previous crop was corn. On November 8, 2021, wheat cultivars P25R40 and P25R61 were drilled at 7.5 inches spacing. All fungicide applications were applied at 15 gal/acre and 40 psi using a CO_2 backpack sprayer equipped with a 10-foot boom, fitted with six TJ-VS 8002 nozzles spaced 20 inches apart and directed forward and backward at a 45-degree angle. Fungicides were applied on May 24 at Feekes growth stage 10.5.1. All plots were inoculated with a mixture of isolates of *Fusarium graminearum* endemic to Indiana on May 25. The spore suspension (50,000 spores/ml) was applied at 300 ml/plot with the CO_2 handheld sprayer. Disease ratings were assessed on June 13, 2022. Fusarium head blight (FHB) incidence was measured as the number of infected heads out of 60 plants in each plot and calculated as a percentage. FHB severity was rated by visually assessing the percentage of the infected head. The FHB index was calculated as (% FHB incidence multiplied by average FHB severity)/100 per plot. Values for each plot were averaged before analysis. The eight center rows of each plot were harvested with a Kincaid plot combine on July 7, and yields were adjusted to 13.5% moisture. All data were analyzed in SAS 9.4 (SAS Institute, Cary, NC). A generalized linear mixed model analysis of variance was performed using PROC GLIMMIX. Values are least squares means, and values with different letters are significantly different based on a least squares means test ($\alpha = 0.05$).

In 2022, weather conditions were moderately favorable for FHB. FHB was the most prominent disease. P25R61 had reduced FHB incidence, FHB severity, deoxynivalenol (DON), percentage of FDK, and yield as compared to P25R40. FHB incidence and index were reduced by all fungicides over the nontreated, inoculated control on June 13 (Table 18). The concentration of DON was reduced by all fungicides applied at 10.5.1 over the nontreated controls. There was no difference in treatment for FHB severity, percentage of fusarium damaged kernels (FDK), and yield over nontreated control.

TABLE 18. *Effect of Cultivar and Fungicide on Fusarium Head Blight (FHB), Deoxynivalenol (DON), Fusarium Damaged Kernels (FDK), and Yield of Wheat*

CULTIVAR OR TREATMENT AND RATE/ACRE[z]	FHB DI[y] %	FHB DS[x] %	FHB INDEX[w]	FDK[v] %	DON[u] PPM	YIELD[t] BU/ACRE
Cultivar						
P25R40 (scab susceptible)	13.5 a[s]	12.8 a	1.6 a	9.6 a	0.783 a[s]	64.0 a
P25R61 (scab resistant)	4.0 b	5.3 b	0.3 b	8.9 b	0.229 b	59.0 b
Fungicide rate/Acre						
Nontreated control, inoculated	15.0 a	11.8	2.0 a	9.5	0.770 a	59.4
Prosaro 421 SC 6.5 fl oz at 10.5.1	8.8 b	6.5	0.9 b	8.9	0.398 b	60.3
Miravis Ace 275 SC 13.7 fl oz at 10.5.1	5.8 b	12.8	0.5 b	9.5	0.350 b	65.6
Prosaro Pro 400 SC 10.3 fl oz at 10.5.1	6.0 b	7.6	0.5 b	9.3	0.409 b	64.1
Sphaerex 7.3 fl oz at 10.5.1	9.0 b	5.5	0.7 b	8.6	0.311 b	59.6
Nontreated control, noninoculated	7.7 b	10.3	0.9 b	9.8	0.796 a	60.1
P-value *cultivar*[s]	0.0001	0.0074	0.0001	0.0360	0.0001	0.0256
P-value *fungicide*	0.0019	0.5127	0.0040	0.2881	0.0001	0.4030
P-value *cultivar*fungicide*	0.0009	0.9962	0.0053	0.1449	0.0352	0.3051

[z] Fungicides treatments were applied on May 24 at Feekes growth stage 10.5.1. All treatments contained a nonionic surfactant (Preference) at a rate of 0.125% v/v. All plots were inoculated with *Fusarium graminearum* spore suspension (50,000 spores/ml) after the treatment at Feekes 10.5.1 except nontreated, noninoculated plots. Spore suspension was applied at 300 ml/plot with the handheld sprayer on May 25.

[y] FHB disease incidence (DI) was measured as the number of infected heads out of 60 plants in each plot and calculated as a percentage.

[x] FHB disease severity (DS) was rated by visually assessing the percentage of the infected head. FHB = Fusarium head blight.

[w] The FHB index was calculated as (% FHB incidence multiplied by average FHB severity)/100 per plot.

[v] Fusarium damaged kernels (FDK) was visually assessed as a percentage (0–100%) of the infected head.

[u] Analysis of mycotoxin deoxynivalenol (DON) was completed by the University of Minnesota DON Testing Lab.

[t] Yields were adjusted to 13.5% moisture after harvest on July 7.

[s] All data were analyzed in SAS 9.4 (SAS Institute, Cary, NC). A generalized linear mixed model analysis of variance was performed using PROC GLIMMIX. Values are least squares means, and values with different letters are significantly different based on a least squares means test (α = 0.05).

EVALUATION OF FOLIAR FUNGICIDES FOR WHEAT DISEASE MANAGEMENT IN CENTRAL INDIANA, 2022 (WHT22-06.ACRE)

I. Miranda, S. Shim, and D. E. P. Telenko, Department of Botany and Plant Pathology, Purdue University West Lafayette, IN 47907-2054

WHEAT (*TRITICUM AESTIVUM* P25R40)

Fusarium head blight, *Fusarium graminearum*

A trial was established at the Purdue Agronomy Center for Research and Education (ACRE) in Tippecanoe County, Indiana. The experiment was a randomized complete block design with four replications. Plots were 7.5 feet wide and 20 feet long and consisted of 12 rows spaced 7.5 inches apart, and the center of each plot was used for evaluation. The previous crop was corn. On November 8, 2021, wheat cultivar P25R40 was drilled at 7.5 inches spacing. All fungicide applications were applied at 15 gal/acre and 40 psi using a CO_2 backpack sprayer equipped with a 10-foot boom, fitted with six TJ-VS 8002 nozzles spaced 20 inches apart and directed forward and backward at a 45-degree angle. Fungicides were applied on May 24 at Feekes growth stage 10.5.1. All plots were inoculated with a mixture of isolates of *Fusarium graminearum* endemic to Indiana on May 25. The spore suspension (50,000 spores/ml) was applied at 300 ml/plot with the CO_2 handheld sprayer. Disease ratings were assessed on June 13. Fusarium head blight (FHB) incidence was measured as the number of infected heads out of 60 plants in each plot and calculated as a percentage. FHB severity was rated by visually assessing the percentage of the infected head. The FHB index was calculated as (% FHB incidence multiplied by average FHB severity)/100 per plot. Values for each plot were averaged before analysis. The eight center rows of each plot were harvested with a Kincaid plot combine on July 7, and yields were adjusted to 13.5% moisture. All data were analyzed in SAS 9.4 (SAS Institute, Cary, NC). A generalized linear mixed model analysis of variance was performed using PROC GLIMMIX. Values are least squares means, and values with different letters are significantly different based on a least squares means test (α = 0.05).

In 2022, weather conditions were moderately favorable for FHB. FHB was the most prominent disease. FHB incidence and index were not significantly reduced by all fungicide treatments over the nontreated control (Table 19). No difference in FHB severity was detected compared to the nontreated control, but Prosaro at 8.2 fl oz had significantly higher severity as compared to all other fungicide treatments. The concentration of deoxynivalenol (DON) was significantly reduced in all treatments over the nontreated control except Prosaro at 6.5 fl oz/acre. There were no significant differences in percentage of fusarium kernels damage (FDK) and yield of wheat.

TABLE 19. *Effect of Cultivar and Fungicide on Fusarium Head Blight (FHB), Deoxynivalenol (DON), Fusarium Damaged Kernels (FDK), and Yield of Wheat*

TREATMENT AND RATE/ACRE[z]	FHB % DI[y]	FHB % DS[x]	FHB INDEX[w]	DON[v] PPM	FDK[u] %	YIELD[t] BU/ACRE
Nontreated control	30.8	2.9 ab	0.9	1.310 a	7.5	65.6
Prosaro 421 SC 8.2 fl oz	20.4	4.0 a	0.8	0.698 bc	7.5	70.6
Prosaro Pro 400 SC 10.3 fl oz	27.1	1.9 b	0.5	0.455 c	8.5	65.9
Miravis Ace 5.2 SC 13.7 fl oz	20.9	2.1 b	0.5	0.678 bc	9.0	68.8
Sphaerex 7.3 fl oz	17.1	2.3 b	0.4	0.603 bc	9.5	70.5
Prosaro 421 SC 6.5 fl oz	27.5	2.3 b	0.7	0.978 ab	9.5	67.1
P-value[s]	0.2175	0.0296	0.0875	0.0071	0.6385	0.0976

[z] Fungicides were applied on May 24 at Feekes growth stage 10.5.1 and contained a nonionic surfactant at a rate of 0.125% v/v. All plots were inoculated with a mixture of isolates of *Fusarium graminearum* endemic to Indiana on May 25, with a spore suspension (50,000 spores/ml) applied at 300 ml/plot.

[y] FHB disease incidence (DI) was measured as the number of infected heads out of 60 plants in each plot and calculated as a percentage on July 13.

[x] FHB disease severity (DS) was rated by visually assessing the percentage of the infected head on July 13.

[w] The FHB index was calculated as (% FHB incidence multiplied by average FHB severity)/100 per plot. FHB = Fusarium head blight.

[v] Analysis of mycotoxin deoxynivalenol (DON) was completed by the University of Minnesota DON Testing Lab.

[u] Fusarium damaged kernels (FDK) was visually assessed as a percentage (0–100%) of the infected heads.

[t] Yields were adjusted to 13.5% moisture after harvest on July 7.

[s] All data were analyzed in SAS 9.4 (SAS Institute, Cary, NC). A generalized linear mixed model analysis of variance was performed using PROC GLIMMIX. Values are least squares means, and values with different letters are significantly different based on a least squares means test (α = 0.05).

EVALUATION OF FOLIAR FUNGICIDES FOR WHEAT IN CENTRAL INDIANA, 2022 (WHT22-08.ACRE)

I. Miranda, S. Shim, and D. E. P. Telenko, Department of Botany and Plant Pathology, Purdue University West Lafayette, IN 47907-2054

WHEAT (*TRITICUM AESTIVUM* P25R40)

Leaf blotch, *Septoria tritici, Stagonospora nodorum*

A trial was established at the Purdue Agronomy Center for Research and Education (ACRE) in Tippecanoe County, Indiana. The experiment was a randomized complete block design with four replications. Plots were 7.5 feet wide and 20 feet long and consisted of 12 rows spaced 7.5 inches apart, and the center of each plot was used for evaluation. The previous crop was corn. On November 13, 2021, wheat cultivar P25R40 was drilled at 7.5 inches spacing. All fungicide applications were applied at 15 gal/acre and 40 psi using a CO_2 backpack sprayer equipped with a 10-foot boom, fitted with six TJ-VS 8002 nozzles spaced 20 inches apart and directed forward and backward at a 45-degree angle. Foliar fungicides were applied on May 12 at Feekes growth stage 8 and on May 24 at Feekes growth stage 10.5.1. Leaf blotch severity was assessed on June 13 and was rated by visually assessing the percentage of symptomatic leaf tissue on the flag leaf on five plants per plot and then averaged. The eight center rows of each plot were harvested with a Kincaid plot combine on July 7, and yields were adjusted to 13.5% moisture. All data were analyzed in SAS 9.4 (SAS Institute, Cary, NC). A generalized linear mixed model analysis of variance was performed using PROC GLIMMIX. Values are least squares means, and values with different letters are significantly different based on a least squares means test ($\alpha = 0.05$).

In 2022, weather conditions were moderately favorable for FHB, and very little leaf blotch developed. No differences were detected for FHB incidence, severity, and index as compared to the nontreated control (data not presented). No differences were detected for leaf spot severity or wheat yield.

TABLE 20. *Effect of Cultivar and Fungicide on Leaf Blotch Severity and Yield of Wheat*

TREATMENT AND RATE/ACRE[z]	LEAF BLOTCH SEVERITY[y] %	YIELD[x] BU/ACRE
Nontreated control	0.6	54.7
Nexicor EC 7.0 fl oz at Feekes 8	1.0	57.3
Topguard EQ 10.0 fl oz at Feekes 8	0.0	52.5
Priaxor 4.0 fl oz at Feekes 8	0.4	60.1
Trivapro SE 9.4 fl oz at Feekes 8	0.1	55.0
Delaro 325 SC 8.0 fl oz at Feekes 8	0.0	58.0
Quilt Xcel 2.2 SE 10.5 fl oz at Feekes 8	0.3	53.9
Tilt 3.6 EG 4.0 fl oz at Feekes 8	0.0	52.1
Headline SC 6.0 fl oz at Feekes 8	0.0	57.1
Prosaro 421 SC 6.5 fl oz at Feekes 10.5.1	0.3	57.6
P-value[s]	0.4598	0.5547

[z] Fungicides were applied on May 24 at Feekes growth stages 8 and 10.5.1 and contained a nonionic surfactant at a rate of 0.125% v/v.

[y] Leaf blotch severity was visually assessed as a percentage of symptomatic leaf tissue on flag leaf on June 13. Five plants per plot were assessed, and data were averaged.

[x] Yields were adjusted to 13.5% moisture after harvest on July 7.

[s] All data were analyzed in SAS 9.4 (SAS Institute, Cary, NC). A generalized linear mixed model analysis of variance was performed using PROC GLIMMIX. Values are least squares means, and values with different letters are significantly different based on a least squares means test ($\alpha = 0.05$).

PINNEY PURDUE AGRICULTURAL CENTER (PPAC)

UNIFORM FUNGICIDE COMPARISON FOR TAR SPOT IN CORN IN NORTHWESTERN INDIANA, 2022 (COR22-02_UFTTAR.PPAC)

M. S. Mizuno, S. Shim, and D. E. P. Telenko, Department of Botany and Plant Pathology, Purdue University West Lafayette, IN 47907-2054

CORN (*ZEA MAYS* W2585VT2P)

Tar spot, *Phyllachora maydis*

A trial was established at the Pinney Purdue Agricultural Center (PPAC) in Porter County, Indiana. The experiment was a randomized complete block design with four replications. Plots were 10 feet wide and 30 feet long and consisted of four rows, and the two center rows were used for evaluation. The previous crop was corn. Standard practices for grain corn production in Indiana were followed. Corn hybrid W2585VT2P was planted in 30-inch row spacing at a rate of 34,000 seeds/acre on May 31. The field was overhead irrigated at 1 inch once a week unless weekly rainfall was 1 inch or more to encourage disease. All foliar fungicide applications were applied at 15 gal/acre and 40 psi using a Lee self-propelled sprayer equipped with a 10-foot boom, fitted with six TJ-VS 8002 nozzles spaced 20 inches apart. Fungicides were applied on August 2 at silk (R1) and on August 23 at dough (R4) growth stages and three weeks after treatments. Disease ratings were assessed on September 20 and October 5 at dent (R5) and maturity (R6) growth stages, respectively. Tar spot was rated by visually assessing the percentage of stromata per leaf (0–100%) at ear leaf on five plants in each plot. Canopy greenness was rated by visually assessing the percentage (0–100%) of the whole plot for crop canopy that remained green at maturity (R6) growth stage. The two center rows of each plot were harvested on November 4, and yields were adjusted to 15.5% moisture All disease and yield data were analyzed using a mixed model analysis of variance. Values are least squares means, and values with different letters are significantly different based on a least squares means test (α = 0.05).

In 2022 weather conditions were not favorable for disease, and very little disease developed in plots. All treatments reduced tar spot stromata severity over the nontreated control except for Miravis Neo on September 20 at R5 (Table 21). Tar spot stromata severity was significantly reduced over the nontreated control by all fungicide treatments except Aproach Prima, Miravis Neo, and Delaro Complete on October 5 at R6. Head-line AMP followed by Velytma, Aproach Prima followed by Headline AMP, Miravis Neo followed by Head-line AMP, Headline AMP followed by Delaro Complete at 3 weeks after treatment, and a single application of Aproach Prima significantly increased the percentage of canopy green over the nontreated control on Oc-tober 5 at R6. There was no significant effect of treatment on yield of corn.

TABLE 21. *Effect of Fungicide Programs on Tar Spot Severity, Canopy Greenness, and Yield of Corn*

TREATMENT, RATE/ACRE, AND TIMING[z]	TAR SPOT %[y] SEPTEMBER 20	TAR SPOT %[y] OCTOBER 5	CANOPY GREEN[x] %	YIELD[w] BU/ACRE
Nontreated control	0.07 b	0.45 a	13.8 e	200.5
Veltyma 3.34 S 7.0 fl oz at R1	0.00 c	0.11 c	32.5 a-e	205.5
Aproach Prima 2.34 SC 6.8 fl oz at R1	0.03 c	0.39 a	42.5 ab	220.1
Miravis Neo 2.5 SE 13.7 fl oz at R1	0.12 a	0.39 a	25.0 b-e	216.0
Delaro Complete 458 SC 8.0 fl oz at R1	0.02 c	0.31 ab	37.5 a-e	216.0
Headline AMP 1.68 SC 10.0 fl oz at R1	0.01 c	0.12 bc	22.5 cde	207.7
Veltyma 3.34 S 7.0 fl oz at R1 fb Headline AMP 1.68 SC 10.0 fl oz at 3 WAT	0.00 c	0.01 c	30.0 a-e	206.2
Aproach Prima 2.34 SC 6.8 fl oz at R1 fb Headline AMP 1.68 SC 10.0 fl oz at 3 WAT	0.01 c	0.06 c	42.5 ab	213.5
Miravis Neo 2.5 SE 13.7 fl oz at R1 fb Headline AMP 1.68 SC 10.0 fl oz at 3 WAT	0.01 c	0.06 c	40.0 abc	211.7
Delaro Complete 458 SC 8 fl oz at R1 fb Headline AMP 1.68 SC 10.0 fl oz at 3 WAT	0.00 c	0.04 c	30.0 a-e	209.8
Headline AMP 1.68 SC 10.0 fl oz at R1 fb Veltyma 3.34 S 7.0 fl oz at 3 WAT	0.00 c	0.02 c	47.5 a	216.5
Headline AMP 1.68 SC 10.0 fl oz at R1 fb Aproach Prima 2.34 SC 6.8 fl oz at 3WAT	0.00 c	0.09 c	20.0 de	210.5
Headline AMP 1.68 SC 10.0 fl oz at R1 fb Miravis Neo 2.5 SE 13.7 fl oz at 3 WAT	0.00 c	0.05 c	17.5 e	204.6
Headline AMP 1.68 SC 10.0 fl oz at R1 fb Delaro Complete 458 SC 8.0 fl oz at 3 WAT	0.00 c	0.04 c	42.5 ab	222.3
Headline AMP 1.68 SC 10.0 fl oz at R1 fb Headline AMP 1.68 SC 10.0 fl oz at 3 WAT	0.00 c	0.10 c	32.5 a-e	189.7
P-value[v]	*0.0001*	*0.0001*	*0.0143*	*0.5419*

[z] Fungicides were applied on August 2 at silk (R1) corn growth stage and on August 23 three weeks after treatment (WAT) at dough (R4). All treatments applied contained a nonionic surfactant (Preference) at a rate of 0.25% v/v. fb = followed by.

[y] Tar spot stromata severity was visually assessed as a percentage (0–100%) of leaf area on five plants in each plot at the ear leaf on September 20 at R5 and on October 5 at R6.

[x] Canopy greenness was visually assessed as a percentage (0–100%) on October 5 at R6.

[w] Yields were adjusted to 15.5% moisture after harvest on November 4.

[v] All data were analyzed in SAS 9.4 (SAS Institute, Cary, NC). A generalized linear mixed model analysis of variance was performed using PROC GLIMMIX. Values are least squares means, and values with different letters are significantly different based on a least squares means test ($\alpha = 0.05$).

EVALUATION OF HYBRIDS AND FUNGICIDE TIMING FOR TAR SPOT IN CORN IN NORTHWESTERN INDIANA, 2022 (COR22-03.PPAC)

K. M. Goodnight, S. Shim, and D. E. P. Telenko, Department of Botany and Plant Pathology, Purdue University West Lafayette, IN 47907-2054

CORN (*ZEA MAYS* W2585VT2P AND P0589AMX)

Tar spot, *Phyllachora maydis*

A trial was established at the Pinney Purdue Agricultural Center (PPAC) in Porter County, Indiana. The experiment was a randomized complete block design with four replications. Plots were 10 feet wide and 30 feet long and consisted of four rows, and the two center rows were used for evaluation. The previous crop was corn. Standard practices for grain corn production in Indiana were followed. Corn hybrids W2585VT2P and P0589AMXT were planted in 30-inch row spacing at a rate of 34,000 seeds/acre on May 20. All fungicide applications were applied at 15 gal/acre and 40 psi using a Lee self-propelled sprayer equipped with a 10-foot boom, fitted with six TJ-VS 8002 nozzles spaced 20 inches apart. Delaro Complete fungicide was applied on July 14, July 21, August 2, August 12, and August 23 at eight-leaf (V8), 10-leaf (V10), tassel/silk (VT/R1), blister (R2), and dough (R4) growth stages, respectively. A weather-based prediction model for tar spot (https://connect.doit.wisc.edu/cpn-risk-tool/) was used, and applications were made at the V8 and VT/R1 growth stages. Disease ratings were assessed on September 21 and on October 5 at the dent (R5) and maturity (R6) growth stages, respectively. Tar spot was rated by visually assessing the percentage of stromata (0–100%) per leaf on five plants in each plot at the ear leaf. Values for each plot were averaged before analysis. Canopy green was rated by visually assessing the percentage (0–100%) of the whole plot for crop canopy that remained green at the dent (R5) and maturity (R6) growth stages. The two center rows of each plot were harvested on October 20, and yields were adjusted to 15.5% moisture. All disease and yield data were analyzed using a mixed model analysis of variance (SAS 9.4, 2019). Values are least squares means, and values with different letters are significantly different based on a least squares means test (α = 0.05).

In 2022, weather conditions were not favorable for tar spot disease. Tar spot was the most prominent disease in the trial and reached low severity. There was no significant interaction between hybrid and fungicide for disease and yield; therefore, main effects are presented. No differences between cultivars were observed on September 21. Tar spot stromata severity was significantly reduced with the tar spot tolerant cultivar compared to the tar spot susceptible cultivar on October 5 (Table 22). No differences between fungicide treatments and nontreated control were observed for tar spot on September 21. On October 5, Delaro Complete applied at R2 significantly reduced the severity of tar spot over the nontreated control but was not significantly different from application at R4 or when applied using the tar spot model (V8 followed by VT/R1). Percent canopy green significantly increased for the susceptible cultivar compared to the tolerant cultivar on October 5, but no significant differences between cultivars for canopy greenness were observed on September 21. The susceptible cultivar had significantly higher moisture than the tolerant cultivar. Delaro Complete applied at VT/R1 significantly increased canopy greenness compared to nontreated control on September 21. No differences between treatments for canopy greenness on October 5 were observed. There was no significant difference between hybrids and fungicide applications and the nontreated control for grain moisture and yield.

TABLE 22. *Effect of Fungicide on Tar Spot Severity, Canopy Greenness, and Yield of Corn*

TREATMENT, RATE/ACRE, AND TIMING[z]	TAR SPOT[y] % SEPTEMBER 21	TAR SPOT[y] % OCTOBER 5	CANOPY GREEN[x] % SEPTEMBER 21	CANOPY GREEN[x] % OCTOBER 5	HARVEST MOISTURE %	TEST WEIGHT LB/BU	YIELD[w] BU/ ACRE
Hybrids							
Susceptible (W2583VT2P)	0.03	0.15 a	52.9	20.0 a	22.7 a	54.1	212.4
Tolerant (P0589AMXT)	0.01	0.03 b	53.5	8.0 b	20.5 b	55.0	206.1
Fungicide Programs							
Nontreated control	0.02	0.11 ab	48.8 bc	10.9	21.2	54.1	202.8
Delaro Complete 458 SC 8.0 fl oz at V10	0.07	0.17 a	44.4 c	10.8	21.1	54.4	212.9
Delaro Complete 458 SC 8.0 fl oz at VT/R1	0.02	0.14 a	61.9 a	15.6	21.8	54.0	202.9
Delaro Complete 458 SC 8.0 fl oz at R2	0.00	0.02 c	55.6 ab	16.3	21.6	56.5	212.5
Delaro Complete 458 SC 8.0 fl oz at R4	0.01	0.02 bc	49.4 bc	12.0	21.6	54.2	216.2
Delaro Complete 458 SC 8.0 fl oz at tar spot model	0.01	0.10 abc	59.4 ab	18.3	22.6	54.2	208.1
P-value *hybrid*[v]	0.1337	0.0001	0.8448	0.0006	0.0001	0.0311	0.1051
P-value *fungicide*	0.2530	0.0049	0.0218	0.6465	0.0733	0.0088	0.2351
P-value *hybrid*fungicide*	0.3248	0.0621	0.6855	0.7098	0.5455	0.0144	0.2030

[z] Fungicide treatments were applied on July 14, July 21, August 2, August 12, and August 23 at eight-leaf (V8), 10-leaf (V10), tassel/silk (VT/R1), blister (R2), and dough (R4) growth stages, respectively. tar spot model applications were made on July 14 and August 2. fb = followed by.

[y] Tar spot stromata was visually assessed as a percentage (0–100%) of leaf area on five plants in each plot at the ear leaf on September 21 and on October 5.

[x] Canopy greeness was visually assessed as a percentage (0–100%) on September 21 and on October 5.

[w] Yields were adjusted to 15.5% moisture after harvest on October 20.

[v] All data were analyzed in SAS 9.4. A mixed model analysis of variance was performed using PROC GLIMMIX. Values are least squares means, and values with different letters are significantly different based on a least squares means test (α = 0.05).

EVALUATION OF PRODUCTS AND HYBRIDS FOR TAR SPOT IN ORGANIC CORN IN NORTHWESTERN INDIANA, 2022 (COR22-04.PPAC)

C. R. Da Silva, S. Shim, and D. E. P. Telenko, Department of Botany and Plant Pathology, Purdue University West Lafayette, IN 47907-2054

CORN (*ZEA MAYS* ALSEED O.84-95UP AND O.52-96)

Tar spot, *Phyllachora maydis*
Gray leaf spot, *Cercospora zeae-maydis*
Northern corn leaf blight, *Exserohilum turcicum*

A trial was established at the Pinney Purdue Agricultural Center (PPAC) in Porter County, Indiana. The experiment was a randomized complete block design with four replications. Plots were 10 feet wide and 30 feet long and consisted of four rows, and the two center rows were used for evaluation. The previous crop was corn. Standard practices for organic grain corn production in Indiana were followed. Organic hybrids ALSEED O.84-95UP and O.52-96 were planted in 30-inch row spacing at a rate of 34,000 seeds/acre on May 20. The field was overhead irrigated weekly at 1 inch unless weekly rainfall was 1 inch or higher to encourage disease. All fungicide applications were applied at 15 gal/acre and 40 psi using a Lee self-propelled sprayer equipped with a 10-foot boom, fitted with six TJ-VS 8002 nozzles spaced 20 inches apart. Fungicide treatments were applied on July 26 at silk (R1) growth stage. Disease ratings were assessed on August 29 at dent (R5) growth stage. Tar spot, gray leaf spot (GLS), and northern corn leaf blight (NCLB) were rated by visually assessing the percentage (0–100%) per leaf area on five plants in each plot at the ear leaf. Values for the five leaves were averaged before analysis. Canopy greenness was rated by visually assessing the percentage (0–100%) of the whole plot for crop canopy that remained green at dent (R5) growth stage. The two center rows of each plot were harvested on October 21, and yields were adjusted to 15.5% moisture. All disease and yield data were analyzed using a mixed model analysis of variance, and means were separated using least squares means test ($\alpha = 0.05$).

In 2022 weather conditions were not favorable for disease, and very little disease developed in plots. Tar spot, GLS, and NCLB were present in the trial but only remained at low levels. There was no significant interaction between hybrid and fungicide; therefore, main effects of hybrid and fungicide were evaluated (Table 23). No differences between hybrids for tar spot, GLS, and NCLB were detected. ALSEED o.84-95UP had significantly greener canopy and higher yield than O.52-96 hybrid. There were no differences between treatments and the nontreated control for tar spot, GLS, NCLB, canopy greenness, and yield of corn.

TABLE 23. *Effect of Fungicide on Foliar Disease Severity, Canopy Greenness, and Yield of Corn*

TREATMENT AND RATE/ACRE[z]	TAR SPOT[y] %	GLS[y] %	NCLB[y] %	CANOPY[x] GREEN %	YIELD[w] BU/A
Hybrids					
ALSEED O.84-95UP	0.001	0.03	0.0	86.4 a	201.0 a
O.52-96	0.004	0.04	0.1	66.7 b	186.5 b
Fungicide programs					
Nontreated control	0.013	0.02	0.0	73.8	200.1
Headline AMP 1.68 SE 10 fl oz	0.000	0.05	0.0	78.8	206.6
Serifel WP 16 fl oz	0.003	0.06	0.0	76.6	202.0
Actinovate AG 12 ox	0.000	0.04	0.0	73.8	183.1
Badge X2 SC 1.8 lb	0.000	0.03	0.3	75.6	176.5
OxiDate 5.0 128 fl oz	0.000	0.03	0.0	80.6	194.1
P-value hybrid[v]	*0.4303*	*0.8665*	*0.3246*	*0.0001*	*0.0395*
P-value fungicide	*0.4583*	*0.9611*	*0.4331*	*0.7011*	*0.1048*
*P-value hybrid*fungicide*	*0.3710*	*0.2080*	*0.4331*	*0.5945*	*0.0234*

[z] Fungicide treatments were applied on July 26 at silk (R1) growth stage.

[y] Tar spot stromata, GLS, and NCLB were visually assessed as a percentage (0–100%) of leaf area on five plants in each plot at the ear leaf on September 29. GLS = gray leaf spot, NCLB = northern corn leaf blight.

[x] Canopy greenness was visually assessed as a percentage (0–100%) of green on September 29.

[w] Yields were adjusted to 15.5% moisture after harvest on October 21.

[v] All data were analyzed in SAS 9.4 (SAS Institute, Cary, NC). A generalized linear mixed model analysis of variance was performed using PROC GLIMMIX. Values are least squares means, and values with different letters are significantly different based on a least squares means test (α = 0.05).

FUNGICIDE EFFICACY AND TIMING FOR TAR SPOT IN CORN IN NORTHWESTERN INDIANA, 2022 (COR22-05.PPAC)

C. R. Da Silva, S. Shim, and D. E. P. Telenko, Department of Botany and Plant Pathology, Purdue University West Lafayette, IN 47907-2054

CORN (*ZEA MAYS* W2585VT2P)

Tar spot, *Phyllachora maydis*

A trial was established at the Pinney Purdue Agricultural Center (PPAC) in Porter County, Indiana. The experiment was a randomized complete block design with four replications. Plots were 10 feet wide and 30 feet long and consisted of four rows, and the two center rows were used for evaluation. The previous crop was corn. Standard practices for grain corn production in Indiana were followed. Corn hybrid W2585VT2P was planted in 30-inch row spacing at a rate of 34,000 seeds/acre on May 31. The field was overhead irrigated weekly at 1 inch unless weekly rainfall was 1 inch or higher to encourage disease. All fungicide applications were applied at 15 gal/acre and 40 psi using a Lee self-propelled sprayer equipped with a 10-foot boom, fitted with six TJ-VS 8002 nozzles spaced 20 inches apart. Fungicides were applied on September 9 at first detection of tar spot, on July 14 at V8 growth stage, on August 2 at tassel (VT) growth stage, on August 19 at milk (R3) growth stage, at first detection of tar spot fb 3 weeks after treatment (WAT) (this timing not receive application due to PHI of fungicide), on August 2 at V8 fb 3 WAT, on August 23 at VT fb 3 WAT, and on September 9 at R3 fb 3 WAT. Disease ratings were assessed on September 20 at dent (R5) and on October 5 at maturity (R6) growth stages. Tar spot was rated by visually assessing the percentage of stromata per leaf on five plants in each plot at the ear leaf. Values for the five leaves were averaged before analysis. Percent of greenness was rated by visually assessing the percentage (0–100%) of the whole plot for crop canopy that remained green at dent (R5) and maturity (R6) growth stages. The two center rows of each plot were harvested on November 4, and yields were adjusted to 15.5% moisture. All data were analyzed in SAS 9.4. A generalized linear mixed model analysis of variance was performed using PROC GLIMMIX. Values are least squares means, and values with different letters are significantly different based on a least squares means test ($\alpha = 0.05$).

In 2022 weather conditions were not favorable for disease, and very little disease developed in plots. Tar spot was present in the trial but remained at low levels. All fungicide programs significantly reduced the percentage of stromata of tar spot over the nontreated control 1 at the dent (R5) growth stage (Table 24). There were no significant differences between treatments for canopy greenness on September 20. All application timings of Veltyma significantly reduced the severity of stromata of tar spot over nontreated controls on October 5 at the R6 growth stage except the applications at V8 and at first detection followed by 3 WAT. The application at first detection (September 9) was the only timing of Lucento that reduced tar spot over nontreated controls on October 5 at R6. All fungicide treatments significantly increased the percent of canopy greenness over the nontreated control 1 at R6 except Veltyma at R3, V8 followed by 3 WAT, and Lucento at first detection of tar spot at R3, first detection followed by 3 WAT, V8 followed by 3 WAT, VT followed by 3 WAT, and R3 followed by 3 WAT. No significant differences in yield were detected.

TABLE 24. *Effect of Fungicide on Tar Spot Severity, Canopy Greenness, and Yield of Corn*

TREATMENT, RATE/ACRE, AND TIMING[z]	TAR SPOT[y] % SEPTEMBER 20	CANOPY GREEN[x] % SEPTEMBER 20	TAR SPOT[y] % OCTOBER 5	CANOPY GREEN[x] % OCTOBER 5	YIELD[w] BU/ACRE
Nontreated control 1	0.8 a	72.5	1.9 a	41.3 fg	223.6
Veltyma 3.34 S 7.0 fl oz at FD	0.0 b	87.5	0.1 d	63.8 abc	229.2
Veltyma 3.34 S 7.0 fl oz at V8	0.0 b	86.3	0.5 cd	62.5 abc	231.7
Veltyma 3.34 S 7.0 fl oz at VT	0.0 b	85.0	0.2 d	70.0 a	236.5
Veltyma 3.34 S 7.0 fl oz at R3	0.0 b	76.0	0.1 d	53.0 b-f	248.1
Veltyma 3.34 S 7.0 fl oz at FD fb 3 WAT	0.2 b	80.0	1.9 a	58.8 a-d	247.6
Veltyma 3.34 S 7.0 fl oz at V8 fb 3 WAT	0.0 b	81.7	0.0 d	50.0 c-g	259.2
Veltyma 3.34 S 7.0 fl oz at VT fb 3 WAT	0.1 b	85.0	0.0 d	65.0 ab	254.3
Veltyma 3.34 S 7.0 fl oz at R3 fb 3 WAT	0.0 b	88.8	0.0 d	57.5 a-e	234.3
Nontreated control 2	0.2 b	76.3	0.9 bc	57.5 a-e	235.9
Lucento 7.17 SC 5.0 fl oz at FD	0.0 b	87.5	0.1 d	52.5 b-f	225.1
Lucento 7.17 SC 5.0 fl oz at V8	0.0 b	82.0	1.3 ab	57.5 a-d	247.4
Lucento 7.17 SC 5.0 fl oz at VT	0.1 b	80.0	0.6 cd	61.3 abc	252.0
Lucento 7.17 SC 5.0 fl oz at R3	0.0 b	77.5	0.5 cd	43.8 efg	227.5
Lucento 7.17 SC 5.0 fl oz at FD fb 3 WAT	0.0 b	73.8	0.4 cd	36.3 g	243.7
Lucento 7.17 SC 5.0 fl oz at V8 fb 3 WAT	0.2 b	73.8	0.6 bcd	42.5 fg	226.0
Lucento 7.17 SC 5.0 fl oz at VT fb 3 WAT	0.1 b	76.3	0.6 bcd	45.0 d-g	235.9
Lucento 7.17 SC 5.0 fl oz at R3 fb 3 WAT	0.1 b	83.8	0.2 d	52.5 b-f	256.6
P-value[v]	0.0347	0.0617	0.0001	0.0005	0.0921

[z] Fungicide treatments were applied on September 9 at first detection (FD) of tar spot, on July 14 at V8 growth stage, on August 2 at tassel (VT) growth stage, on August 19 at milk (R3) growth stage, at first detection of tar spot fb 3 weeks after treatment (WAT) (this timing not receive application due to PHI of fungicide), on August 2 at V8 fb 3 WAT, on August 23 at VT fb 3 WAT, and on September 9 at R3 fb 3 WAT. All treatments at VT or later contained a nonionic surfactant at a rate of 0.25% v/v. fb = followed by.

[y] Tar spot stromata was visually assessed as a percentage (0–100%) of leaf area on five plants in each plot at the ear leaf on September 20 and October 5.

[x] Canopy greenness was visually assessed as a percentage (0–100%) of green on October 5. Yields were adjusted to 15.5% moisture after harvest on November 4.

[v] All data were analyzed in SAS 9.4. A mixed model analysis of variance was performed using PROC GLIMMIX. Values are least squares means, and values with different letters are significantly different based on least square difference test (α = 0.05).

EVALUATION OF XYWAY PROGRAMS FOR TAR SPOT CONTROL IN NORTHWESTERN INDIANA, 2022 (COR22-14.PPAC)

I. L. Miranda, S. Shim, and D. E. P. Telenko, Department of Botany and Plant Pathology, Purdue University West Lafayette, IN 47907-2054

CORN (*ZEA MAYS* W2585VT2P)

Tar spot, *Phyllachora maydis*

A trial was established at the Pinney Purdue Agricultural Center (PPAC) in Porter County, Indiana. The experiment was a randomized complete block design with four replications. Plots were 10 feet wide and 30 feet long and consisted of four rows, and the two center rows were used for evaluation. The previous crop was corn. Standard practices for grain corn production in Indiana were followed. Corn hybrid W2585VT2P was planted in 30-inches row spacing at a rate of 2 seeds/foot on May 20. Standard practices for nonirrigated grain corn production in Indiana were followed. Xyway was applied in-furrow and 2x0 at planting at 10 gal/acre. All foliar fungicide applications were applied at 15 gal/acre and 40 psi using a Lee self-propelled sprayer equipped with a 10-foot boom, fitted with six TJ-VS 8002 nozzles spaced 20 inches apart, and were applied on August 2 at tassel/silk (VT/R1) growth stage. Disease ratings were assessed on September 21 at dent (R5) and October 3 at maturity (R6) growth stages. Tar spot severity was rated by visually assessing the percentage (0–100%) of stromata per leaf on five plants in each plot at the ear leaf. Values for each plot were averaged before analysis. The two center rows of each plot were harvested on October 20, and yields were adjusted to 15.5% moisture. All data were analyzed using a mixed model analysis of variance (SAS 9.4). Values are least squares means, and values with the same letter are not significantly different based on a least squares means test ($\alpha = 0.05$).

In 2022 weather conditions were not favorable for disease, and very little disease developed in plots. Tar spot was present in the trial but remained at low levels. There were no significant differences between treatments and the nontreated control for severity of tar spot stromata (Table 25). There were no significant differences between treatments and nontreated plots for canopy greenness, harvest moisture, test weight, and yield of corn.

TABLE 25. *Effect of Fungicide on Tar Spot Severity, Canopy Greenness, and Yield of Corn*

TREATMENT, RATE/ACRE, AND APPLICATION[z]	TAR SPOT %[y] SEPTEMBER 21	TAR SPOT %[y] OCTOBER 3	CANOPY GREEN[x] %	HARVEST MOISTURE %	TEST WEIGHT LB/BU	YIELD[w] BU/ACRE
Nontreated control	0.02	0.11	10.0	21.1	55.4	197.3
Xyway LFR 15.2 fl oz in-furrow	0.02	0.22	7.5	20.3	55.6	194.8
Xyway LFR 15.2 fl oz at 2xo	0.02	0.29	8.8	20.9	55.3	195.4
Xyway LFR 8.35 fl oz in-furrow fb Topguard EQ 5.0 fl oz at VT/R1	0.02	0.20	6.3	20.2	56.0	196.4
Topguard EQ 5.0 fl oz at VT/R1	0.01	0.15	10.0	20.6	56.4	197.4
Veltyma 7.0 fl oz at VT/R1	0.01	0.24	8.8	19.1	56.6	208.9
P-value[v]	0.5522	0.1013	0.7279	0.5781	0.6473	0.6418

[z] Xyway was applied in-furrow and 2xo at planting at 10 gal/acre on May 20. Foliar fungicides were applied on August 2 at the tassel/silk (VT/R1) growth stage. fb = followed by.

[y] Tar spot stromata severity was visually assessed as the percentage (0–100%) of leaf area covered by stromata on five plants in each plot at the ear leaf at the dent (R5) on September 21 and maturity (R6) growth stages on October 3.

[x] Canopy greenness was visually assessed as a percentage (0–100%) on October 3.

[w] Yields were adjusted to 15.5% moisture after harvest on October 20.

[v] All data were analyzed in SAS 9.4 (SAS Institute, Cary, NC). A generalized linear mixed model analysis of variance was performed using PROC GLIMMIX. Values are least squares means, and values with different letters are significantly different based on a least squares means test (α = 0.05).

FUNGICIDE COMPARISON FOR TAR SPOT IN CORN IN NORTHWESTERN INDIANA, 2022 (COR22-16.PPAC)

C. R. Da Silva, S. Shim, and D. E. P. Telenko, Department of Botany and Plant Pathology, Purdue University West Lafayette, IN 47907-2054

CORN (*ZEA MAYS* W2585VT2P)

Tar spot, *Phyllachora maydis*

A trial was established at the Pinney Purdue Agricultural Center (PPAC) in Porter County, Indiana. The experiment was a randomized complete block design with four replications. Plots were 10 feet wide and 30 feet long and consisted of four rows, and the two center rows were used for evaluation. The previous crop was corn. Standard practices for grain corn production in Indiana were followed. Corn hybrid W2585VT2P was planted in 30-inch row spacing at a rate of 34,000 seeds/acre on May 31. All fungicide applications were applied at 15 gal/acre and 40 psi using a Lee self-propelled sprayer equipped with a 10-foot boom, fitted with six TJ-VS 8002 nozzles spaced 20 inches apart. Fungicides were applied on August 2 at the tassel/silk (VT/R1) growth stage. Disease ratings were assessed on September 19 and October 5 at early dent (R5) and late dent (R5) growth stages, respectively. Tar spot was rated by visually assessing the percentage of stromata per leaf on five plants as a percentage (0–100%) of leaf area on five plants in each plot at the ear leaf. Values for the five leaves were averaged before analysis. Canopy greenness was rated by visually assessing the percentage (0–100%) of the whole plot for crop canopy that remained green at the dent (R5) and maturity (R6) growth stages. The two center rows of each plot were harvested on November 4, and yields were adjusted to 15.5% moisture. All disease and yield data were analyzed using a mixed model analysis of variance, and means were separated using Fisher's least significant difference ($\alpha = 0.05$).

In 2022 weather conditions were not favorable for disease, and very little disease developed in plots. Tar spot was present in the trial but remained at low levels. There was no significant effect of treatment on tar spot stromata over the nontreated controls on September 19 and October 5 (Table 26). There was no significant effect of treatment on canopy greenness, harvest moisture, test weight, and yield or corn.

TABLE 26. *Effect of Fungicide on Tar Spot Severity, Canopy Greenness, and Yield of Corn*

TREATMENT AND RATE/ACRE[z]	TAR SPOT[y] % SEPTEMBER 19	TAR SPOT[y] % OCTOBER 5	CANOPY GREEN[x] %	HARVEST MOISTURE %	TEST WEIGHT LB/BU	YIELD[w] BU/ACRE
Nontreated control	0.1	0.9	53.8	19.1	55.7	218.4
Miravis Neo 2.5 SE 13.7 fl oz	0.2	1.1	57.5	20.0	55.4	217.5
Delaro Complete 458 SC 8.0 fl oz	0.0	0.8	61.3	20.4	54.5	215.1
Veltyma 3.24 S 7.0 fl oz	0.1	0.7	56.3	20.8	54.8	215.0
BioMineral Exp A 7.6 fl oz	0.2	1.6	52.5	19.7	54.5	218.7
BioMineral Exp B 56.3 fl oz	0.3	1.4	43.8	19.5	55.1	217.2
Brixen 10.0 fl oz	0.2	1.2	56.3	20.0	56.3	204.8
Brixen 10.0 fl oz + Stilo PSR 5.0 fl oz	0.1	1.5	51.3	19.7	55.1	214.6
Brixen at 13.0 fl oz	0.2	1.0	61.3	20.3	57.5	216.6
Brixen 13.0 fl oz + Stilo PSR 5.0 fl oz	0.2	1.1	56.3	19.8	55.7	214.3
Quilt Xcel 2.2 SE 10.5 fl oz	0.2	1.6	61.3	20.4	55.2	207.5
Quilt Xcel 2.2 SE 10.5 fl oz + Stilo PSR 5.0 fl oz	0.2	1.0	56.3	19.7	55.2	215.3
Headline Amp 1.68 SC 10.0 fl oz	0.0	0.3	55.0	19.2	55.5	219.0
P-value[v]	0.7474	0.6699	0.5121	0.7253	0.8623	0.6582

[z] Fungicide treatments were applied at the tassel/silk (VT/R1) grow stage on August 2. All treatments contained a nonionic surfactant at a rate of 0.25% v/v except Biomineral Exp A and B.

[y] Tar spot stromata was visually assessed as a percentage (0–100%) of leaf area on five plants in each plot at the ear leaf on September 19.

[x] Canopy greenness was visually assessed as a percentage (0–100%) of crop canopy on September 20 and October 5.

[w] Yields were adjusted to 15.5% moisture after harvest on November 4.

[v] All data were analyzed in SAS 9.4 (SAS Institute, Cary, NC). A generalized linear mixed model analysis of variance was performed using PROC GLIMMIX. Values are least squares means, and values with different letters are significantly different based on a least squares means test (α = 0.05).

EVALUATION OF FUNGICIDE PROGRAMS FOR TAR SPOT IN CORN IN NORTHWESTERN INDIANA, 2022 (COR22-18.PPAC)

S. Shim and D. E. P. Telenko, Department of Botany and Plant Pathology, Purdue University
West Lafayette, IN 47907-2054

CORN (*ZEA MAYS* W2585VT2P)

Tar spot, *Phyllachora maydis*

A trial was established at the Pinney Purdue Agricultural Center (PPAC) in Porter County, Indiana. The experiment was a randomized complete block design with four replications. Plots were 10 feet wide and 30 feet long and consisted of four rows, and the two center rows were used for evaluation. The previous crop was corn. Standard practices for grain corn production in Indiana were followed. Corn hybrid W2585VT2P was planted in 30-inch row spacing at a rate of 2 seeds/foot on May 20. All foliar fungicide applications were applied at 15 gal/acre and 40 psi using a Lee self-propelled sprayer equipped with a 10-foot boom, fitted with six TJ-VS 8002 nozzles spaced 20 inches apart. Fungicides were applied on July 14, August 2, August 12, August 19, and August 23 at 8-leaf (V8), tassel/silk (VT/R1), blister (R2), milk (R3), and dough (R4) growth stages, respectively. Tar spot model applications were made on July 14 and August 2. Disease ratings were assessed on September 21 and October 5 at dent (R5) and maturity (R6) growth stages, respectively. Tar spot was rated by visually assessing the percentage of stromata per leaf on five plants in each plot at the ear leaf. Values for each plot were averaged before analysis. The two center rows of each plot were harvested on November 3, and yields were adjusted to 15.5% moisture. All data were analyzed in SAS 9.4 (SAS Institute, Cary, NC). A generalized linear mixed model analysis of variance was performed using PROC GLIMMIX. Values are least squares means, and values with different letters are significantly different based on a least squares means test (α = 0.05).

In 2022, weather conditions were not favorable for tar spot disease. Tar spot was the most prominent disease in the trial and reached low severity. No significant differences between treatments and nontreated control were detected for tar spot stromata severity on September 21 and October 5 (Table 27). Aproach Prima applied at R2 and R4 increased canopy greenness over the nontreated control on October 5. There were no significant differences for harvest moisture, test weight, and yield of corn.

TABLE 27. *Effect of Fungicide on Tar Spot Severity, Canopy Greenness, and Yield of Corn*

TREATMENT, RATE/ACRE, AND TIMING[z]	TAR SPOT[y] % SEPTEMBER 21	TAR SPOT[y] % OCTOBER 5	CANOPY GREEN[x] %	HARVEST MOISTURE %	TEST WEIGHT LB/BU	YIELD[w] BU/ACRE
Nontreated control	0.04	0.9	17.5 b	20.3	53.8	205.0
Aproach Prima 2.34 SC 6.8 fl oz at VT/R1	0.00	0.4	21.3 b	20.7	53.7	216.0
Aproach Prima 2.34 SC 6.8 fl oz at R2	0.00	0.3	43.8 a	18.5	54.2	210.4
Aproach Prima 2.34 SC 6.8 fl oz at R3	0.00	0.4	12.5 b	20.5	53.6	198.6
Aproach Prima 2.34 SC 6.8 fl oz at R4	0.03	0.6	43.8 a	20.4	54.0	202.0
Aproach Prima 2.34 SC 6.8 fl oz at VT/R1 fb Trivapro 2.21 SE 13.7 fl oz at R3/R4	0.00	0.2	12.5 b	21.0	52.9	205.4
Aproach Prima 2.34 SC 6.8 fl oz at VT/R1 fb Delaro Complete 485 SC 9.0 fl oz at R3/R4	0.00	0.1	13.8 b	19.6	54.4	210.2
Aproach Prima 2.34 SC 6.8 fl oz at VT/R1 fb Headline AMP 1.68 SC 10.0 fl oz at R3/R4	0.00	0.1	26.3 ab	20.1	54.5	207.7
Aproach Prima 2.34 SC 6.8 fl oz at Tar spot model	0.00	0.2	25.0 b	19.5	55.2	211.5
P-value[v]	0.2628	0.3460	0.0050	0.5233	0.3880	0.5431

[z] Fungicides were applied on July 14, August 2, August 12, August 19, and August 23 at 8-leaf (V8), tassel/silk (VT/R1), blister (R2), milk (R3), and dough (R4) growth stages, respectively. Tar spot model application were made on July 14 and August 2. fb = followed by.

[y] Tar spot stromata was visually assessed as a percentage (0–100%) of leaf area on five plants in each plot at the ear leaf on September 21 and October 5.

[x] Canopy greenness was visually assessed as a percentage (0–100%) on October 5.

[w] Yields were adjusted to 15.5% moisture after harvest on November 3.

[v] All data were analyzed in SAS 9.4 (SAS Institute, Cary, NC). A generalized linear mixed model analysis of variance was performed using PROC GLIMMIX. Values are least squares means, and values with different letters are significantly different based on a least squares means test ($\alpha = 0.05$).

FUNGICIDE COMPARISON FOR FOLIAR DISEASES IN CORN IN NORTHWESTERN INDIANA, 2022 (COR22-24.PPAC)

S. Shim and D. E. P. Telenko, Department of Botany and Plant Pathology, Purdue University
West Lafayette, IN 47907-2054

CORN (*ZEA MAYS* W2585VT2P)

Tar spot, *Phyllachora maydis*

A trial was established at the Pinney Purdue Agricultural Center (PPAC) in Porter County, Indiana. The experiment was a randomized complete block design with four replications. Plots were 10 feet wide and 30 feet long and consisted of four rows, and the two center rows were used for evaluation. The previous crop was corn. Standard practices for grain corn production in Indiana were followed. Corn hybrid W2585VT2P was planted in 30-inch row spacing at a rate of 34,000 seeds/acre on May 31. All foliar fungicide applications were applied at 15 gal/acre and 40 psi using a Lee self-propelled sprayer equipped with a 10-foot boom, fitted with six TJ-VS 8002 nozzles spaced 20 inches apart. Fungicides were applied on July 14, July 26, August 2, August 12, and August 19 at V8, V12-V14, silk (R1), blister (R2), and milk (R3) growth stages. Tar spot model applications were made on July 14 and August 2 at V8 and silk (R1) growth stages, respectively. Disease ratings were assessed on September 21 and October 3 at dent (R5) and maturity (R6) growth stages, respectively. Tar spot was rated by visually assessing the percentage of stromata per leaf on five plants in each plot at the ear leaf. Values for each plot were averaged before analysis. The two center rows of each plot were harvested on November 3, and yields were adjusted to 15.5% moisture. All data were analyzed in SAS 9.4 (SAS Institute, Cary, NC). A generalized linear mixed model analysis of variance was performed using PROC GLIMMIX. Values are least squares means, and values with different letters are significantly different based on a least squares means test ($\alpha = 0.05$).

In 2022, weather conditions were not favorable for tar spot. Tar spot was present in the trial but remained at low levels. There was no significant effect of fungicide treatment on tar spot on September 21 (Table 28). On October 3, Veltyma at R1, Delaro Complete at R2, Miravis Neo applied at R2 and R3, and the programs with multiple applications significantly reduced tar spot over the nontreated control. There were no significant differences between treatments and the nontreated control for canopy greenness, harvest moisture, test weight, and yield of corn.

TABLE 28. *Effect of Fungicide Treatment on Tar Spot Severity, Canopy Greenness, and Yield of Corn*

TREATMENT, RATE/ACRE, AND TIMING[z]	TAR SPOT[y] % SEPTEMBER 21	TAR SPOT[y] % OCTOBER 3	CANOPY GREEN[x] %	HARVEST MOISTURE %	TEST WEIGHT LB/BU	YIELD[w] BU/ACRE
Nontreated control	0.11	0.39 ab	25.0	18.4	55.1	201.6
Miravis Neo 2.5 SE 13.7 fl oz at V12	0.08	0.30 abc	30.0	19.3	54.5	205.1
Trivapro 2.21 SE 13.7 fl oz at V12	0.05	0.22 bcd	32.5	19.0	55.2	203.6
Miravis Neo 2.5 SE 13.7 fl oz at R1	0.05	0.41 a	22.5	18.9	54.5	206.1
Trivapro 2.21 SE 13.7 fl oz at R1	0.05	0.25 a-d	32.5	19.9	54.2	208.9
Veltyma 3.34 S 7.0 fl oz at R1	0.02	0.11 de	32.5	20.6	53.7	200.0
Delaro Complete 458 SC 8.0 fl oz at R1	0.02	0.09 de	30.0	18.3	55.9	199.9
Miravis Neo 2.5 SE 13.7 fl oz at R2	0.01	0.10 de	27.5	18.4	55.4	201.8
Miravis Neo 2.5 SE 13.7 fl oz at R3	0.02	0.14 cde	32.5	19.7	60.0	206.4
Miravis Neo 2.5 SE 13.7 fl oz at V12 fb Miravis Neo 2.5 SE 13.7 fl oz at R2	0.03	0.14 cde	32.5	20.3	53.8	205.1
Veltyma 3.34 S 7.0 fl oz at R1 fb Veltyma 3.34 S 7.0 fl oz at R3	0.00	0.03 e	35.0	19.5	54.7	210.0
Miravis Neo 2.5 SE 13.7 fl oz at R1 fb Miravis Neo 2.5 SE 13.7 fl oz at R3	0.03	0.14 cde	27.5	19.6	54.4	208.1
Miravis Neo 2.5 SE 13.7 fl oz at Tar spot model (V8 and R1)	0.04	0.22 bcd	21.3	19.5	54.7	204.7
P-value[v]	*0.0574*	*0.0016*	*0.7648*	*0.3369*	*0.4255*	*0.4817*

[z] Foliar fungicides were applied on July 14, July 26, August 2, August 12, and August 19 at V8, V12–V14, silk (R1), blister (R2), and milk (R3) growth stages, respectively. Tar spot model applications were made on July 14 and August 2 at V8 and silk (R1) growth stages, respectively. Fungicide treatments contained a nonionic surfactant at a rate of 0.25% v/v except V8 and V12 applications. fb = followed by.

[y] Tar spot stromata was visually assessed as a percentage (0–100%) of leaf area on five plants in each plot at the ear leaf on September 21 and October 3.

[x] Canopy greenness was visually assessed as a percentage (0–100%) of crop canopy green on October 3.

[w] Yields were adjusted to 15.5% moisture after harvest on November 3.

[v] All data were analyzed in SAS 9.4 (SAS Institute, Cary, NC). A generalized linear mixed model analysis of variance was performed using PROC GLIMMIX. Values are least squares means, and values with different letters are significantly different based on a least squares means test (α = 0.05).

EVALUATION OF XYWAY PROGRAMS IN CORN FOR TAR SPOT IN NORTHWESTERN INDIANA, 2022 (COR22-27.PPAC)

S. Shim and D. E. P. Telenko, Department of Botany and Plant Pathology, Purdue University
West Lafayette, IN 47907-2054

CORN (*ZEA MAYS* W2585VT2P)

Tar spot, *Phyllachora maydis*

A trial was established at the Pinney Purdue Agricultural Center (PPAC) in Porter County, Indiana. The experiment was a randomized complete block design with four replications. Plots were 10 feet wide and 30 feet long and consisted of four rows, and the two center rows were used for evaluation. The previous crop was corn. Standard practices for grain corn production in Indiana were followed. Corn hybrid W2585VT20 was planted in 30-inch row spacing at a rate of 34,000/acre on May 20. Xyway applications were applied 2xo at 10 gal/acre on May 23 by a CO_2 backpack sprayer. All foliar fungicide applications were applied at 15 gal/acre and 40 psi using a Lee self-propelled sprayer equipped with a 10-foot boom, fitted with six TJ-VS 8002 nozzles spaced 20 inches apart. Fungicides were applied on July 21, August 2 and August 12 at V10, silk (R1), and blister (R2) growth stages, respectively. Disease ratings were assessed on September 21 and October 3 at dent (R5) and maturity (R6) growth stages, respectively. Tar spot was rated by visually assessing the percentage of stromata per leaf on five plants in each plot at the ear leaf. Values for each plot were averaged before analysis. Canopy greenness was visually assessed as a percentage (0–100%) of crop canopy green on October 3. The two center rows of each plot were harvested on October 20, and yields were adjusted to 15.5% moisture. All data were analyzed in SAS 9.4 (SAS Institute, Cary, NC). A generalized linear mixed model analysis of variance was performed using PROC GLIMMIX. Values are least squares means, and values with different letters are significantly different based on a least squares means test ($\alpha = 0.05$).

In 2022, weather conditions were not favorable for disease. Tar spot was the most prominent diseases in the trial and reached low severity. All treatments reduced tar spot over the nontreated control on September 20 (Table 29). On October 3, no treatments were significantly different from the nontreated control. Xyway LFR 9.5 fl oz was applied 2xo followed by Adastrio 7.0 fl oz at R2, Xyway LFR at 15.2 fl oz applied 2xo followed by Veltyma 7.0 fl oz at R2, and Delaro 5.0 at V10 followed by Delaro Complete at R2 had a significantly greener canopy on October 20 as compared to the nontreated control. There was no significant effect of treatments on harvest moisture, test weight, and yield of corn.

TABLE 29. *Effect of Fungicide on Tar Spot Severity, Canopy Greenness, and Yield of Corn*

TREATMENT, RATE/ACRE, AND TIMING[z]	TAR SPOT[y] % SEPTEMBER 20	TAR SPOT[y] % OCTOBER 3	CANOPY GREEN[x] %	HARVEST MOISTURE %	TEST WEIGHT LB/BU	YIELD[w] BU/ ACRE
Nontreated control	0.12 a	0.11 a–d	1.5 d	21.4	54.3	188.6
Xyway LFR 9.5 fl oz at plant 2xo	0.03 bc	0.15 ab	1.5 d	21.8	54.0	192.5
Xyway LFR 15.2 fl oz at plant 2xo	0.05 bc	0.16 a	2.8 bcd	20.1	55.0	201.8
Xyway LFR 9.5 fl oz at plant 2xo fb Adastrio 4.0 SC 7.0 fl oz at R1	0.06 b	0.13 abc	0.7 d	21.2	54.3	195.2
Xyway LFR 9.5 fl oz at plant 2xo fb Adastrio 4.0 SC 7.0 fl ox at R2	0.01 c	0.05 cd	7.8 a	18.8	55.2	207.3
Xyway LFR 15.2 fl oz at plant 2xo fb Veltyma 3.34 S 7.0 fl oz R2	0.00 c	0.03 d	7.3 ab	20.9	54.5	198.5
Topguard EQ 4.29 10 fl oz at V10 fb Adastrio 4.0 SC 7.0 fl oz at R2	0.01 bc	0.06 cd	3.0 bcd	20.3	55.8	200.0
Adastrio 4.0 SC 7.0 fl oz at R1	0.05 bc	0.16 a	2.5 cd	21.0	54.3	197.6
Delaro 325 SC 5.0 fl oz at V10 fb Delaro Complete 458 SC 8.0 fl oz at R2	0.00 c	0.04 cd	7.0 abc	19.0	53.6	209.1
Veltyma 3.34 S 7.0 fl oz at R1	0.01 c	0.07 bcd	2.5 cd	20.7	54.9	199.6
P-value[v]	*0.0007*	*0.0150*	*0.0307*	*0.1616*	*0.6507*	*0.3633*

[z] Xyway applications were applied 2xo at 10 gal/acre on May 23. Foliar fungicides were applied on July 21, August 2, and August 12 at the V10, silk (R1), and blister (R2) growth stages, respectively. fb = followed by.

[y] Tar spot stromata was visually assessed as a percentage (0–100%) of leaf area on five plants in each plot at the ear leaf on September 20 and October 3.

[x] Canopy greenness was visually assessed as a percentage (0–100%) of crop canopy green on October 3.

[w] Yields were adjusted to 15.5% moisture after harvest on October 20.

[v] All data were analyzed in SAS 9.4 (SAS Institute, Cary, NC). A generalized linear mixed model analysis of variance was performed using PROC GLIMMIX. Values are least squares means, and values with different letters are significantly different based on a least squares means test (α = 0.05).

EVALUATION OF FUNGICIDE TIMING AND APPLICATION FOR TAR SPOT IN CORN IN NORTHWESTERN INDIANA, 2022 (COR22-29.PPAC)

S. Shim and D. E. P. Telenko, Department of Botany and Plant Pathology, Purdue University
West Lafayette, IN 47907-2054

CORN (*ZEA MAYS* W2585VT2P)

Tar spot, *Phyllachora maydis*

A trial was established at the Pinney Purdue Agricultural Center (PPAC) in Porter County, Indiana. The experiment was a randomized complete block design with four replications. Plots were 10 feet wide and 30 feet long and consisted of four rows, and the two center rows were used for evaluation. The previous crop was corn. Standard practices for grain corn production in Indiana were followed. Corn hybrid W2585VT2P was planted in 30-inch row spacing at a rate of 34,000 seeds/acre on May 31. All foliar fungicide applications were applied at 15 gal/acre and 40 psi using a Lee self-propelled sprayer equipped with a 10-foot boom, fitted with six TJ-VS 8002 nozzles spaced 20 inches apart. Fungicides were applied on July 14, July 21, August 2, and August 19 at the V8, V10, tassel/silk (VT/R1), and R3 growth stages, respectively. Tar spot model applications were made at V8 and tassel/silk (VT/R1) growth stages. Disease ratings were assessed on September 19 and October 5 at the dent (R5) and dent/maturity (R5/R6) growth stages, respectively. Tar spot was rated by visually assessing the percentage of stromata per leaf on five plants in each plot at the ear leaf. Values for each plot were averaged before analysis. The two center rows of each plot were harvested on November 3, and yields were adjusted to 15.5% moisture. All disease and yield data were analyzed using a mixed model analysis of variance, and means were separated using Fisher's least significant difference ($\alpha = 0.05$).

In 2022, weather conditions were not favorable for tar spot disease. Tar spot was the most prominent diseases in the trial and reached low severity. All fungicide treatments reduced tar spot stromata severity except Delaro Complete applied at V10 on October 3 (Table 30). No significant differences were detected for canopy greenness, harvest moisture, and test weight. No treatments significantly increased yield over the nontreated control.

TABLE 30. *Effect of Fungicide on Tar Spot Severity, Canopy Greenness, and Yield of Corn*

TREATMENT, RATE/ACRE, AND TIMING[z]	TAR SPOT[y] %	CANOPY GREEN[x] %	HARVEST MOISTURE %	TEST WEIGHT LB/BU	YIELD[w] BU/ACRE
Nontreated control	0.12 bc	6.8	18.5	54.8	202.5 abc
Delaro Complete 458 SC 8.0 fl oz at V10	0.35 a	7.5	18.8	54.2	208.4 a
Delaro Complete 458 SC 8.0 fl oz at V10 fb Delaro Complete 458 SC 8.0 fl oz at VT/R1	0.04 bc	7.5	18.3	55.2	194.2 cd
Delaro Complete 458 SC 8.0 fl oz at V10 fb Delaro Complete 458 SC 8.0 fl oz at R3	0.01 c	8.8	18.4	54.9	204.3 ab
Veltyma 3.34 S 7.0 fl oz at V10	0.16 b	12.5	18.2	55.8	202.9 abc
Veltyma 3.34 S 7.0 fl oz at V10 fb Veltyma 3.34 S 7.0 fl oz at VT/R1	0.03 c	11.3	19.6	55.7	203.1 abc
Delaro Complete 458 SC 8.0 fl oz at VT/R1	0.07 bc	7.5	18.8	54.8	201.1 abc
Delaro Complete 458 SC 10.0 fl oz at VT/R1	0.04 bc	5.8	19.4	54.7	196.0 bcd
Delaro Complete 458 SC 8.0 fl oz at VT/R1 fb Delaro Complete 458 SC 8.0 fl oz at R3	0.01 c	8.8	18.6	55.2	203.5 abc
Veltyma 3.34 S 7.0 fl oz at VT/R1 fb Veltyma 3.34 S 7.0 fl oz at R3	0.01 c	11.3	18.6	54.7	208.5 a
Miravis Neo 2.5 SE 13.7 fl oz at VT/R1 fb Miravis Neo 2.5 SE 13.7 fl oz at R3	0.10 bc	10.0	19.3	55.3	205.5 a
Delaro 325 SC 4.0 fl oz at V8 fb Delaro 325 SC 8.0 fl oz at VT/R1	0.04 bc	8.8	19.8	54.3	191.6 d
Delaro 325 SC 10.0 fl oz at VT/R1	0.08 bc	9.5	18.9	54.7	202.7 abc
Delaro Complete 458 SC 10.0 fl oz at VT/R1	0.04 bc	7.5	18.2	55.4	202.1 abc
Delaro Complete 458 SC 8.0 fl oz at Tar spot model	0.06 bc	8.3	19.0	54.9	205.7 a
P-value[v]	0.0005	0.5327	0.4253	0.3974	0.0265

[z] Fungicides were applied on August 2 and August 19 at tassel/silk (VT/R1) and milk (R3) growth stages, respectively. Tar spot model applications were made at V8 and tassel/silk (VT/R1). All treatments were applied at VT/R1 or R3 contained a nonionic surfactant (Preference) at a rate of 0.25% v/v. fb = followed by.

[y] Tar spot stromata was visually assessed as a percentage (0–100%) of leaf area on five plants in each plot at the ear leaf on October 3.

[x] Canopy greenness was visually assessed as a percentage (0–100%) of crop canopy green on October 3.

[w] Yields were adjusted to 15.5% moisture after harvest on November 3.

[v] All data were analyzed in SAS 9.4 (SAS Institute, Cary, NC). A generalized linear mixed model analysis of variance was performed using PROC GLIMMIX. Values are least squares means, and values with different letters are significantly different based on a least squares means test (α = 0.05).

EVALUATION OF EFFICACY OF CX-9032 AND CX-10250 FOR TAR SPOT IN CORN IN NORTHWESTERN INDIANA, 2022 (COR22-30.PPAC)

S. Shim and D. E. P. Telenko, Department of Botany and Plant Pathology, Purdue University
West Lafayette, IN 47907-2054

CORN (*ZEA MAYS* W2585VT2P)

Tar spot, *Phyllachora maydis*

A trial was established at the Pinney Purdue Agricultural Center (PPAC) in Porter County, Indiana. The experiment was a randomized complete block design with four replications. Plots were 10 feet wide and 30 feet long and consisted of four rows, and the two center rows were used for evaluation. The previous crop was corn. Standard practices for grain corn production in Indiana were followed. Corn hybrid W2585VT2P was planted in 30-inch row spacing at a rate of 34,000 seeds/acre on May 31. All foliar fungicide applications were applied at 15 gal/acre and 40 psi using a Lee self-propelled sprayer equipped with a 10-foot boom, fitted with six TJ-VS 8002 nozzles spaced 20 inches apart. Fungicides were applied on August 2 and August 19 at the tassel/silk (VT/R1) and milk (R3) growth stages, respectively. Disease ratings were assessed on September 19 and October 5 at the dent (R5) and dent/maturity (R5/R6) growth stages, respectively. Tar spot was rated by visually assessing the percentage of stromata per leaf on five plants in each plot at the ear leaf. Values for each plot were averaged before analysis. The two center rows of each plot were harvested on November 4, and yields were adjusted to 15.5% moisture. All data were analyzed in SAS 9.4 (SAS Institute, Cary, NC). A generalized linear mixed model analysis of variance was performed using PROC GLIMMIX. Values are least squares means, and values with different letters are significantly different based on a least squares means test (α = 0.05).

In 2022, weather conditions were not favorable for tar spot disease. Tar spot was the most prominent diseases in the trial and reached low severity. No differences were detected between fungicide treatments and the nontreated control for tar spot stromata severity on September 19 and October 3 (Table 31). No differences between treatments and the nontreated control were detected for canopy greenness, harvest moisture, test weight, and yield of corn.

TABLE 31. *Effect of Fungicide on Tar Spot Severity, Canopy Greenness, and Yield of Corn*

TREATMENT, RATE/ACRE, AND TIMING[z]	TAR SPOT[y] % SEPTEMBER 19	TAR SPOT[y] % OCTOBER 3	CANOPY GREEN[x] %	HARVEST MOISTURE %	TEST WEIGHT LB/BU	YIELD[w] BU/ACRE
Nontreated control	0.4	2.2	23.8	19.2	69.9	200.8
CX-9032 1.0 qt at V10 fb CX-9032 1.0 qt at VT/R1	0.3	0.4	33.8	20.1	56.9	209.8
CX-9032 1.0 qt at VT/R1	0.1	0.5	36.3	19.6	56.4	201.9
CX-10250 1.0 fl oz at V10 fb CX-10250 1.0 fl oz at VT/R1	0.2	0.2	30.0	19.7	57.0	204.6
Veltyma 3.34 S 7.0 fl oz at VT/R1	0.0	0.0	42.5	19.8	56.7	203.6
P-value[v]	0.1012	0.3485	0.3723	0.9438	0.5049	0.5576

[z] Fungicides were applied on August 2 and August 19 at the tassel/silk (VT/R1) and milk (R3) growth stages, respectively. All treatments were applied at VT contained a nonionic surfactant (Preference) at a rate of 0.25% v/v. fb = followed by.

[y] Tar spot stromata was visually assessed as a percentage (0–100%) of leaf area on five plants in each plot at the ear leaf on September 19 and October 3.

[x] Canopy greenness was visually assessed as a percentage (0–100%) of crop canopy green on October 5.

[w] Yields were adjusted to 15.5% moisture after harvest on November 4.

[v] All data were analyzed in SAS 9.4 (SAS Institute, Cary, NC). A generalized linear mixed model analysis of variance was performed using PROC GLIMMIX. Values are least squares means, and values with different letters are significantly different based on a least squares means test (α = 0.05).

FUNGICIDE TIMING AND APPLICATION FOR TAR SPOT IN CORN IN NORTHWESTERN INDIANA, 2022 (COR22-32.PPAC)

S. Shim and D. E. P. Telenko, Department of Botany and Plant Pathology, Purdue University
West Lafayette, IN 47907-2054

CORN (*ZEA MAYS* W2585VT2P)

Tar spot, *Phyllachora maydis*

A trial was established at the Pinney Purdue Agricultural Center (PPAC) in Porter County, Indiana. The experiment was a randomized complete block design with four replications. Plots were 10 feet wide and 30 feet long and consisted of four rows, and the two center rows were used for evaluation. The previous crop was corn. Standard practices for grain corn production in Indiana were followed. Corn hybrid W2585VT2P was planted in 30-inch row spacing at a rate of 34,000 seeds/acre on May 31. All foliar fungicide applications were applied at 15 gal/acre and 40 psi using a Lee self-propelled sprayer equipped with a 10-foot boom, fitted with six TJ-VS 8002 nozzles spaced 20 inches apart. Fungicides were applied on July 14 and August 2 at V8 and tassel/silk (VT/R1) growth stages, respectively. Disease ratings were assessed on September 19 and October 5 at the dent (R5) and dent/maturity (R5/R6) growth stages, respectively. Tar spot was rated by visually assessing the percentage of stromata per leaf on five plants in each plot at the ear leaf. Values for each plot were averaged before analysis. The two center rows of each plot were harvested on November 4, and yields were adjusted to 15.5% moisture. All data were analyzed in SAS 9.4 (SAS Institute, Cary, NC). A generalized linear mixed model analysis of variance was performed using PROC GLIMMIX. Values are least squares means, and values with different letters are significantly different based on a least squares means test ($\alpha = 0.05$).

In 2022, weather conditions were not favorable for tar spot disease. Tar spot was the most prominent disease in the trial and reached low severity. All fungicide significantly reduced the severity of tar spot stromata compared to the nontreated control on September 19 except OR-009E at V8 followed by OR-009E at VT/R1 (Table 32). All fungicides significantly reduced the severity of tar spot stromata compared to the nontreated control on October 5 except OR-009E followed by OR-009E and Veltyma at V8. No differences between treatments and the nontreated control were detected for canopy greenness, harvest moisture, and test weight. Delaro Complete + OR-009E 0.4% v/v applied at VT/R1 significantly increased yield over the nontreated control but was not significantly different from Veltyma +OR-009E at V8, Veltyma at VT/R1, Veltyma + OR-009E at VT/R1, or Veltyma +OR-009E at V8 followed by Veltyma +OR-009E at VT/R1.

TABLE 32. *Effect of Fungicide on Tar Spot Severity, Canopy Greenness, and Yield of Corn*

TREATMENT, RATE/ACRE, AND TIMING[z]	TAR SPOT[y] % SEPTEMBER 19	TAR SPOT[y] % OCTOBER 5	CANOPY GREEN[x] %	HARVEST MOISTURE %	TEST WEIGHT LB/BU	YIELD[w] BU/ACRE
Nontreated control	0.14 a	3.2 a	20.0	20.0	54.4	214.1 bc
OR-009E 0.4% v/v at V8 fb OR-009E 0.4% v/v at VT/R1	0.10 ab	2.1 abc	33.8	21.0	54.2	209.9 c
Veltyma 3.34 S 7.0 fl oz at V8	0.08 bc	2.6 ab	33.8	21.3	53.4	208.6 c
Veltyma 3.34 S 7.0 fl oz + OR-009E 0.4% v/v at V8	0.09 bc	1.4 bcd	36.3	19.3	54.9	218.7 abc
Veltyma 3.34 S 7.0 fl oz at VT/R1	0.02 d	0.3 d	41.3	20.5	53.6	217.3 abc
Veltyma 3.34 S 7.0 fl oz + OR-009E 0.4% v/v at VT/R1	0.02 d	0.6 cd	48.8	20.3	53.5	213.0 bc
Veltyma 3.34 S 7.0 fl oz + OR-009E 0.4% v/v at V8 fb Veltyma3.34 S 7.0 fl oz + OR-009E 0.4% v/v VT/R1	0.01 d	0.2 d	47.5	20.0	53.5	221.1 ab
Delaro Complete 458 SC 8.0 fl oz at VT/R1	0.05 cd	0.6 cd	43.8	20.5	54.2	214.7 bc
Delaro Complete 458 SC 8.0 fl oz + OR-009E 0.4% v/v at VT/R1	0.04 d	0.6 cd	37.5	20.6	53.9	226.7 a
P-value[v]	0.0001	0.0030	0.3059	0.7382	0.5935	0.0499

[z] Fungicides were applied on July 14 and August 2 at V8 and tassel/silk (VT/R1) growth stages, respectively. fb = followed by.

[y] Tar spot stromata was visually assessed as a percentage (0–100%) of leaf area on five plants in each plot at the ear leaf on September 19 and October 5.

[x] Canopy greenness was visually assessed as a percentage (0–100%) of crop canopy green on October 5.

[w] Yields were adjusted to 15.5% moisture after harvest on November 4.

[v] All data were analyzed in SAS 9.4 (SAS Institute, Cary, NC). A generalized linear mixed model analysis of variance was performed using PROC GLIMMIX. Values are least squares means, and values with different letters are significantly different based on a least squares means test (α = 0.05).

EVALUATION OF FOLIAR FUNGICIDES IN CORN IN NORTHWESTERN INDIANA, 2022 (COR22-33.PPAC)

S. Shim and D. E. P. Telenko, Department of Botany and Plant Pathology, Purdue University
West Lafayette, IN 47907-2054

CORN (*ZEA MAYS* W2585VT2P)

Tar spot, *Phyllachora maydis*

A trial was established at the Pinney Purdue Agricultural Center (PPAC) in Porter County, Indiana. The experiment was a randomized complete block design with four replications. Plots were 10 feet wide and 30 feet long and consisted of four rows, and the two center rows were used for evaluation. The previous crop was corn. Standard practices for grain corn production in Indiana were followed. Corn hybrid W2585VT2P was planted in 30-inch row spacing at a rate of 34,000 seeds/acre on May 31. All foliar fungicide applications were applied at 15 gal/acre and 40 psi using a Lee self-propelled sprayer equipped with a 10-foot boom, fitted with six TJ-VS 8002 nozzles spaced 20 inches apart. Fungicides were applied on July 1 and August 2 at the V6 and tassel/silk (VT/R1) growth stages, respectively. Disease ratings were assessed on September 21 and October 3 at the dent (R5) and maturity (R6) growth stages, respectively. Tar spot was rated by visually assessing the percentage of stromata per leaf on five plants in each plot at the ear leaf. Values for each plot were averaged before analysis. The two center rows of each plot were harvested on November 2, and yields were adjusted to 15.5% moisture. All disease and yield data were analyzed using a mixed model analysis of variance, and means were separated using least squares means test ($\alpha = 0.05$).

In 2022, weather conditions were not favorable for tar spot disease. Tar spot was the most prominent disease in the trial and reached low severity. There was no significant difference of tar spot severity over the nontreated control on September 21 (Table 33). Domark at V6 followed by Veltyma at VT/R1 and Affiance at VT/R1 significantly reduced tar spot stroma over the nontreated control on October 3. There was no significant effect of treatment on canopy greenness, harvest moisture, test weight, and yield of corn.

TABLE 33. *Effect of Fungicide on Tar Spot Severity, Canopy Greenness, and Yield of Corn*

TREATMENT, RATE/ACRE, AND TIMING[z]	TAR SPOT[y] % SEPTEMBER 21	TAR SPOT[y] % OCTOBER 3	CANOPY GREEN[x] %	HARVEST MOISTURE %	TEST WEIGHT LB/BU	YIELD[w] BU/ACRE
Nontreated control	0.01	0.46 a	20.0	18.8	56.4	190.6
Affiance 1.5 SC 10.0 fl oz + Domark 230 ME 6.0 fl oz at V6	0.01	0.32 ab	22.5	19.5	56.1	195.9
Affiance 1.5 SC 10.0 fl oz at VT/R1	0.01	0.21 bc	20.0	19.3	55.3	199.8
Domark 230 ME 6.0 fl oz at VT/R1	0.25	0.31 ab	22.5	19.5	56.3	203.8
Affiance 1.5 SC 10.0 fl oz at V6 fb Veltyma 3.34 S 7.0 fl oz at VT/R1	0.01	0.25 abc	22.5	19.7	55.7	204.6
Domark 230 ME 6.0 fl oz at V6 fb Veltyma 3.34 S 7.0 fl oz at VT/R1	0.01	0.08 c	21.3	19.4	55.6	192.1
P-value[v]	*0.4642*	*0.0398*	*0.9990*	*0.9174*	*0.8595*	*0.4524*

[z] Fungicides were applied on July 1 and August 2 at the V6 and tassel/silk (VT/R1) growth stages, respectively. fb = followed by.

[y] Tar spot stromata was visually assessed as a percentage (0–100%) of leaf area on five plants in each plot at the ear leaf on September 21 and October 3.

[x] Canopy greenness was visually assessed as a percentage (0–100%) on October 3.

[w] Yields were adjusted to 15.5% moisture after harvest on November 2.

[v] All data were analyzed in SAS 9.4 (SAS Institute, Cary, NC). A generalized linear mixed model analysis of variance was performed using PROC GLIMMIX. Values are least squares means, and values with different letters are significantly different based on a least squares means test (α = 0.05).

EVALUATION OF DRONE APPLICATIONS FOR TAR SPOT IN CORN IN NORTHWESTERN INDIANA, 2022 (COR22-35.PPAC)

M. S. Mizuno, S. Shim, and D. E. P. Telenko, Department of Botany and Plant Pathology, Purdue University West Lafayette, IN 47907-2054

CORN (*ZEA MAYS* W2585VT2P)

Tar spot, *Phyllachora maydis*

A trial was established at the Pinney Purdue Agricultural Center (PPAC) in Porter County, Indiana. The experiment was a randomized complete block design with four replications. Plots were 10 feet wide and 30 feet long and consisted of four rows, and the two center rows were used for evaluation. The previous crop was corn. Standard practices for grain corn production in Indiana were followed. Corn hybrid W2585VT2P was planted in 30-inch row spacing at a rate of 34,000 seeds/acre on May 31. Veltyma 7.0 fl oz/acre was applied at the silk (R1) corn growth stage on August 21 using three different applicators: a Lee self-propelled sprayer equipped with a 10-foot boom, fitted with six TJ-VS 8002 nozzles spaced 20 inches apart. at 3.6 mph; a CO_2 backpack sprayer equipped with a five-foot boom, fitted with four TJ-VS 8002 nozzles spaced 20-inches apart at 3.1 mph applied 15 gal/acre at 40 PSI; and a DJI Agras T10 drone equipped with a 2.1-gal spray tank with a 16.4-foot spray width pattern using four TJ-VS 8002 nozzles spaced apart at 3.1 mph and an application rate of 1.65 gal/acre at 40 PSI. Disease ratings were assessed on September 20 at the dent (R5) growth stage. Tar spot was rated by visually assessing the percentage of stromata per leaf on five plants in each plot at the ear leaf. Values for each plot were averaged before analysis. The two center rows of each plot were harvested on November 3, and yields were adjusted to 15.5% moisture. All disease and yield data were analyzed using a mixed model analysis of variance. Values are least squares means, and values with different letters are significantly different based on a least squares means test (α = 0.05).

In 2022 weather conditions were not favorable for disease development, and very little disease developed in plots. Tar spot was the most prominent disease in the trial. Veltyma sprayed with the ground rig, backpack, and drone significantly reduced tar spot stromata severity over the nontreated control on September 20 (Table 34). There was no significant difference between treatments for canopy greenness, lodging, and yield of corn.

TABLE 34. *Fungicide Application Effect on Tar Spot Severity, Canopy Greenness, Lodging, and Yield of Corn*

APPLICATION EQUIPMENT AND RATE/ACRE[z]	TAR SPOT[y] %	CANOPY GREEN[x] %	LODGING[w] %	HARVEST MOISTURE %	TEST WEIGHT LB/BU	YIELD[v] BU/ACRE
Nontreated control	0.04 a	38.8	1.5	19.5	55.1	211.0
Ground Rig with Veltyma 7.0 fl oz	0.01 b	42.5	0.5	19.4	55.4	221.5
Backpack with Veltyma 7.0 fl oz	0.00 b	35.0	0.2	19.8	53.4	220.4
Drone with Veltyma 7.0 fl oz	0.00 b	45.0	1.0	19.0	54.7	221.7
P-value[u]	0.0121	0.7250	0.6139	0.5801	0.1102	0.3742

[z] Fungicide treatment was applied on August 21 at the tassel/silk (VT/R1) growth stage using a ground rig, a CO_2 backpack, and a drone. All foliar treatments contained a nonionic surfactant at a rate of 0.25% v/v.

[y] Tar spot stromata severity was visually assessed as a percentage (0–100%) of leaf area on five plants in each plot on September 20.

[x] Canopy greenness was visually assessed as a percentage (0–100%) of crop canopy green on October 5.

[w] Lodging was visually assessed as a percentage (0–100%) of lodged stalks when pushed from shoulder height to 45° from vertical on October 5.

[v] Yield were adjusted to 15.5% moisture after harvest on November 3.

[u] All data were analyzed in SAS 9.4 (SAS Institute, Cary, NC). A generalized linear mixed model analysis of variance was performed using PROC GLIMMIX. Values are least squares means, and values with different letters are significantly different based on a least squares means test ($\alpha = 0.05$).

FUNGICIDE EVALUATION FOR WHITE MOLD IN SOYBEAN IN NORTHWESTERN INDIANA, 2022 (SOY22-04.PPAC)

S. Shim and D. E. P. Telenko, Department of Botany and Plant Pathology, Purdue University
West Lafayette, IN 47907-2054

SOYBEAN (*GLYCINE MAX* P29A19E)

White mold, *Sclerotinia sclerotiorum*

A trial was established at the Pinney Purdue Agricultural Center (PPAC) in Porter County, Indiana. The experiment was a randomized complete block design with four replications. Plots were 10 feet wide and 30 feet long and consisted of four rows, and the two center rows were used for evaluation. The previous crop was soybean. Standard practices for soybean production in Indiana were followed. Soybean cultivar P29A19E was planted in 30-inch row spacing at a rate of 140,000 seeds/acre on May 17. Inoculum of *S. sclerotiorum* was applied on the seedbed at 1.25 g/foot at planting. The field was overhead irrigated weekly at 1 inch unless weekly rainfall was 1 inch or higher to encourage disease. All fungicide applications were applied at 15 gal/acre and 40 psi using a CO_2 backpack sprayer equipped with a 10-foot boom, fitted with six TJ-VS 8002 nozzles spaced 20 inches apart. A spray boom with four 360° drop nozzles was used for the 360 undercover treatment. Fungicides were applied on June 23, July 14, July 16, July 29, and August 2 at the V4, beginning bloom (R1), full bloom (R2), and beginning pod (R3) growth stages, respectively. White mold crop risk model application was made on August 2 at the beginning pod (R3) (https://connect.doit.wisc.edu/cpn-risk-tool/). Disease ratings were assessed on September 15 at the maturity (R7/R8) growth stage. White mold disease was assessed by counting the number of plants in each plot with symptoms. Phytoxicity was visually rated on a scale of 0–100% on September 29. The two center rows of each plot were harvested on September 29, and yields were adjusted to 13% moisture. All data were analyzed in SAS 9.4 (SAS Institute, Cary, NC). A generalized linear mixed model analysis of variance was performed using PROC GLIMMIX. Values are least squares means, and values with different letters are significantly different based on a least squares means test (α = 0.05).

In 2022 weather conditions were favorable for disease development, and very little disease developed in plots. White mold was present in the trial but only reached low levels. There were no significant differences between fungicide treatments and the nontreated control for white mold rating on September 15 (Table 35). Cobra at V4, Cobra at V4 followed by Domark at R3, and Cobra at V4 followed by Topsin at R3 significantly increased phytoxicity on September 29. There was no significant effect of treatment on harvest moisture, test weight, and yield of soybean.

TABLE 35. *Effect of Fungicide on White Mold Incidence, Phytoxicity, and Yield of Soybean*

TREATMENT, RATE/ACRE, AND TIMING[z]	WHITE MOLD #/PLOT[y]	PHYTO[x] %	HARVEST MOISTURE %	TEST WEIGHT LB/BU	YIELD[w] BU/ ACRE
Nontreated control	0.0	0.0 c	14.1	57.0	52.6
Endura 70 WDG 8.0 oz at R1 fb Endura 70 WDG 8.0 oz at R3	0.0	1.3 c	13.7	56.7	52.6
Endura 70 WDG 8.0 oz at R3	0.0	1.3 c	13.9	57.3	52.1
Omega 16.0 fl oz at R3 by 360 under cover	0.2	0.0 c	13.9	57.3	56.2
Omega 16.0 fl oz at R3	0.0	1.3 c	13.8	57.0	53.7
Cobra 8.0 fl oz at V4	0.0	15.0 b	13.4	56.9	54.8
Cobra 8.0 fl oz at V4 fb Domark 5.0 fl oz at R3	0.0	31.3 a	13.6	56.9	51.0
Omega 12.0 fl oz at R1 fb Miravis Neo 2.5 SE 13.7 fl oz at R3	0.0	1.3 c	13.7	56.9	52.1
Delaro Complete 458 SC 8.0 fl oz at R3	0.0	0.0 c	13.9	57.0	54.7
Delaro Complete 458 SC 8.0 fl oz at R3 by 360 under cover	0.0	1.3 c	13.6	57.0	53.0
Headsup Seed Treatment	0.0	0.0 c	13.4	56.7	53.4
Headsup Seed Treatment fb Domark 5.0 fl oz at R3	0.0	0.0 c	13.6	57.3	52.7
Miravis Neo 2.5 SE 16.0 fl oz at R3	0.0	0.0 c	13.6	56.9	53.3
Cobra 8.0 fl oz at V4 fb Topsin 4.5 fl oz at R3	0.0	27.5 a	14.5	57.5	49.5
Omega 16.0 fl oz at white mold crop risk model 360 under cover at R3	0.0	0.0 c	13.6	56.9	53.1
Endura 70 WDG 8.0 oz/A at white mold crop risk model at R3	0.0	0.0 c	13.8	56.9	51.3
P-value[v]	*0.4718*	*0.0001*	*0.1210*	*0.8407*	*0.5188*

[z] Fungicides were applied on June 23, July 14, July 16, July 29, and August 2 at the V4, beginning bloom (R1), full bloom (R2), and beginning pod (R3) growth stages, respectively. On August 2 at the R3 growth stage, fungicide was applied by 360 under cover as indicated, and white mold crop risk model applications were made on August 2 at the R3 growth stage. All fungicide treatments contained a nonionic surfactant at a rate of 0.25% v/v except Cobra. All plots were inoculated with *S. sclerotiorum*. fb = followed by.

[y] White mold disease was assessed by counting the number of plants per plots with symptoms on September 15.

[x] Phytoxicity (Phyto) was visually rated on a scale of 0–100% on September 29.

[w] Yields were adjusted to 13% moisture after harvest on October 1.

[v] All data were analyzed in SAS 9.4 (SAS Institute, Cary, NC). A generalized linear mixed model analysis of variance was performed using PROC GLIMMIX. Values are least squares means, and values with different letters are significantly different based on a least squares means test (α = 0.05).

EVALUATION OF DISEASE MANAGEMENT OPTIONS FOR WHITE MOLD IN ORGANIC SOYBEAN IN NORTHWESTERN INDIANA, 2022 (SOY22-06.PPAC)

C. R. Da Silva, S. Shim, and D. E. P. Telenko, Department of Botany and Plant Pathology, Purdue University West Lafayette, IN 47907-2054

SOYBEAN (*GLYCINE MAX* DWIGHT AND MN1410)

White mold, *Sclerotinia sclerotiorum*

A trial was established at the Pinney Purdue Agricultural Center (PPAC) in Porter County, Indiana. The experiment was a split-plot design with four replications. Plots were 6.7 feet wide and 30 feet long and consisted of four rows, and the two center rows were used for evaluation. The previous crop was sunflower. Cereal rye was planted on September 16, 2021, at a rate of 150 lbs/acre and was terminated using either tillage or roller-crimping on May 17. Standard practices for soybean organic production in Indiana were followed. Organic soybean cultivars Dwight and MN1410 were planted in 20-inch row spacing at a rate of 8 seeds/foot on May 17. Inoculum of *S. sclerotiorum* was applied within the seedbed at 1.25 g/foot at planting, and 60 sclerotia per plot were spread between the middle two rows after tillage and before roller-crimping. The field was overhead irrigated weekly at 1 inch unless weekly rainfall was 1 inch or higher to encourage disease. All fungicides applications were applied at 15 gal/acre and 40 psi using a CO_2 backpack sprayer equipped with a 10-foot boom, fitted with four or six TJ-VS 8002 nozzles spaced 20 or 30 inches apart. Fungicides were applied on July 16 at full bloom (R2) growth stage. Disease ratings were assessed on September 1 at full seed (R6) growth stage. White mold disease incidence was assessed by counting the number of plants in each plot with symptoms. For severity, plants were rated according to the following disease category: 0 = no disease, 1 = lateral branches with white mycelium and lesions, 2 = main stem with white mycelium and sclerotia present, and 3 = entire plant wilted/plant death. The disease severity index (DIX) was calculated by DIX = [sum (disease severity score X number of plants)]/[(maximum disease score) × (disease incidence)] × 100. The center rows of each plot were harvested on October 3, and yields were adjusted to 13% moisture. All disease and yield data were analyzed using a generalized linear mixed model analysis of variance performed using PROC GLIMMIX. Values are least squares means, and values with different letters are significantly different based on a least squares means test (α = 0.05).

In 2022 weather conditions were not favorable for tar spot development, and very little disease developed in the plots. White mold was the most prominent disease and reached low severity. There were no significant interactions between cover crop termination, cultivar, and fungicide, but there was a significant interaction between tillage treatment and cultivar (Table 36). White mold incidence and DIX were greatest in the susceptible cultivar, Dwight, under full tillage, while the moderately resistant cultivar MN1410 had significantly less disease when planted in either full-tillage or roller-crimped rye. In addition, planting Dwight in the roller-crimped rye also significantly reduced disease when compared to the full-tillage system. Canopy greenness was highest and defoliation lowest in the Dwight cultivar versus the MN14. No significant differences were found between tillage treatment and cultivars in yield of soybean. There were no significant differences between the fungicide treatments and the nontreated control for white mold, canopy greenness, defoliation, and yield.

TABLE 36. *Effect of Fungicide on White Mold, Canopy Greenness, Defoliation, and Yield of Soybean*

TREATMENT[z]	WHITE MOLD DI[y] %	WHITE MOLD DIX[x] %	CANOPY GREEN[w] %	DEFOLIATION[v] %	YIELD[u] BU/ ACRE
Cover crop/tillage and cultivar					
Full tillage, Dwight	0.5 a	1.4 a	7.3 b	87.9 b	48.4
Full tillage, MN1410	0.0 b	0.0 b	0.0 c	100.0 a	44.2
Roller-crimped rye, Dwight	0.1 b	0.1 b	34.2 a	49.0 c	49.8
Roller-crimped rye, MN1410	0.0 b	0.0 b	0.0 c	96.3 a	45.3
Fungicide and rate/A					
Nontreated control	0.2	0.2	12.8	83.4	45.9
Endura 70 WDG 8.0 fl oz	0.2	0.5	10.0	83.8	46.8
Double Nickel 55 DWG 2.0 qt	0.3	0.7	7.8	86.3	48.1
Serifel WP 16.0 oz	0.1	0.1	10.3	84.4	47.6
Actinovate AG 12.0 oz	0.2	0.4	10.0	82.5	46.7
BotryStop 2.0 lb	0.2	0.5	11.3	79.4	46.3
P-value *till*[t]	0.0398	0.0564	0.0060	0.0012	0.2143
P-value *cultivar*	0.0001	0.0001	0.0001	0.0001	0.0001
P-value *fungicide*	0.2346	0.4245	0.8131	0.8950	0.5177
P-value *till*cultivar*	0.0001	0.0003	0.0001	0.0001	0.8550
P-value *till*fungicide*	0.6609	0.6535	0.7301	0.4674	0.5392
P-value *cultivar*fungicide*	0.2346	0.4245	0.8131	0.7400	0.5264
P-value *till*cultivar*fungicide*	0.6609	0.6535	0.7301	0.1835	0.1194

[z] Fungicide applications were made on July 16 at the full bloom (R2) growth stage. All plots were inoculated with *S. sclerotiorum* at 1.25 g/foot within the seedbed at planting and 60 sclerotia per plot were spread between the middle two rows before roller-crimped and after tillage.

[y] White mold disease incidence (DI) was assessed by counting the number of plants in each plot with symptoms.

[x] The disease severity index (DIX) was calculated as DIX = [sum (disease severity score X number of plants)]/[(maximum disease score) X (disease incidence)] × 100.

[w] Canopy greenness was visually assessed as a percentage (0–100%) of crop canopy green on September 13.

[v] Defoliation was assessed as a percentage of leaf loss in plot.

[u] Yields were adjusted to 13% moisture after harvest on October 3.

[t] All data were analyzed in SAS 9.4. A generalized linear mixed model analysis of variance was performed using PROC GLIMMIX. Values are least squares means, and values with different letters are significantly different based on a least squares means test (α = 0.05).

EVALUATION THE EFFICACY OF SEED TREATMENTS IN SOYBEAN IN NORTHWESTERN INDIANA, 2022 (SOY22-12.PPAC)

S. Shim and D. E. P. Telenko, Department of Botany and Plant Pathology, Purdue University
West Lafayette, IN 47907-2054

SOYBEAN (*GLYCINE MAX* XO3131E)

Sudden death syndrome, *Fusarium virguliforme*

A trial was established at the Pinney Purdue Agricultural Center (PPAC) in Porter County, Indiana. The experiment was a randomized complete block design with four replications. Plots were 10 feet wide and 30 feet long and consisted of four rows, and the two center rows were used for evaluation. The previous crop was corn. Standard practices for soybean production in Indiana were followed. Soybean cultivar XO3131E was planted in 30-inch row spacing at a rate of 8 seeds/foot on May 17. Inoculum of *F. virguliforme* was applied on the seedbed at 1.25 g/foot at planting. The field was overhead irrigated weekly at 1 inch unless weekly rainfall was 1 inch or higher to encourage disease. Seed treatments were applied by the cooperator. Disease ratings were assessed on September 13 at the full seed (R6) growth stage. Sudden death syndrome (SDS) in each plot was rated for disease incidence (DI) and disease severity (DS). DI incidence was percentage of plants with disease symptoms, and DS was rated using a 1–9 scale, where 1 refers to low disease pressure and 9 refers to premature death of the plant. The SDS index was then calculated using the equation DX = (DI x DS)/9. Root rot rating was assessed on August 16 at the full pod to beginning seed (R4/R5) growth stage by visually assessing dark discoloration on roots. The center rows of each plot were harvested on October 3, and yields were adjusted to 13% moisture. All data were analyzed in SAS 9.4 (SAS Institute, Cary, NC). A generalized linear mixed model analysis of variance was performed using PROC GLIMMIX. Values are least squares means, and values with different letters are significantly different based on a least squares means test (α = 0.05).

In 2022 weather conditions were favorable for disease, and very little disease developed in plots. SDS was present in the trial but only reached low levels. ILevo Votivo significantly reduced root rot over base treatment but was not significantly different from ILevo 720, ILevo 720 + Relenva, ILevo 720 + Relenva+ Experimental, or Saltro (Table 37). There were no significant differences between seed treatments for SDS DI, SDS DS, SDS index, harvest moisture, test weight, and yield of soybean.

TABLE 37. *Effect of Seed Treatment on Root Rot, Sudden Death Syndrome (SDS), and Yield of Soybean*

CULTIVAR AND TREATMENT[z]	ROOT ROT[y] %	SDS DI[x]	SDS DS[w]	SDS INDEX[v]	HARVEST MOISTURE %	TEST WEIGHT LB/BU	YIELD[u] BU/ACRE
Base	4.7 ab	26.3	3.8	10.8	13.6	56.2	56.7
ILEVO 720	2.3 bc	21.3	3.0	7.1	14.4	56.1	59.0
ILEVO 720 + Relenya	2.1 bc	22.5	2.8	7.1	13.7	56.1	58.7
ILEVO 720 + Relenya + Experimental	4.2 abc	27.5	3.3	10.4	13.3	56.2	58.9
ILEVO Votivo	1.9 c	27.5	3.8	11.8	13.3	56.7	56.7
Saltro	2.7 bc	22.5	2.8	7.1	13.9	56.0	59.3
CeraMax	6.1 a	28.8	4.0	12.4	13.4	56.3	56.3
P-value[t]	0.0385	0.5684	0.0947	0.2348	0.4357	0.5458	0.6501

[z] Seed treatment of Base = metalaxyl (8 g AI/100 kg seed) + oxathiapiprolin (7.4 g AI/100 kg seed) + prothioconazole (10 g AI/100 kg seed) + penflufen (5 g AI/100 kg seed) + imidacloprid (0.12 mg AI/seed); Illevo 0.15 m ai/seed; Relenya 10 g AI/seed; Experimental 104 ml/100 kg seed; Illevo Votivo 0.21 mg AI/seed; Saltro 0.075 mg AI/seed; and CeraMax 80 ml/100 kg seed. All plots were inoculated with *F. virguliforme*.

[y] Root rot was visually assessed as a percentage (0–100%) of dark discoloration on roots on August 16.

[x] Disease incidence (DI) was visually assessed as a percentage (0–100%) of plants with disease symptoms on September 13.

[w] SDS disease severity (DS) was visually assessed on a scale of 1–9, where 1 = low disease pressure and 9 = premature death of the plant on September 13.

[v] Disease index was calculated as DI*DS/9.

[u] Yields were adjusted to 13% moisture after harvest on October 3.

[t] All data were analyzed in SAS 9.4 (SAS Institute, Cary, NC). A generalized linear mixed model analysis of variance was performed using PROC GLIMMIX. Values are least squares means, and values with different letters are significantly different based on a least squares means test (α = 0.05).

EVALUATION OF FUNGICIDES FOR WHITE MOLD IN SOYBEAN IN NORTHWESTERN INDIANA, 2022 (SOY22-21.PPAC)

C. R. Da Silva, S. Shim, and D. E. P. Telenko, Department of Botany and Plant Pathology, Purdue University West Lafayette, IN 47907-2054

SOYBEAN (*GLYCINE MAX* P29A19E)

White mold, *Sclerotinia sclerotiorum*

A trial was established at the Pinney Purdue Agricultural Center (PPAC) in Porter County, Indiana. The experiment was a randomized complete block design with four replications. Plots were 10 feet wide and 30 feet long and consisted of four rows, and the two center rows were used for evaluation. The previous crop was soybean. Standard practices for soybean production in Indiana were followed. Soybean cultivar P29A19E was planted in 30-inch row spacing at a rate of 140,000/acre on May 17. Inoculum of *S. sclerotiorum* was applied on the seedbed at 1.25 g/foot at planting. The field was overhead irrigated weekly at 1 inch unless weekly rainfall was 1 inch or higher to encourage disease. All fungicide applications were applied at 15 gal/acre and 40 psi using a Lee self-propelled sprayer equipped with a 10-foot boom, fitted with six TJ-VS 8002 nozzles spaced 20 inches apart. Fungicides were applied on July 14, July 16, and July 29 at the beginning bloom (R1), full bloom (R2), and beginning pod (R3) growth stages, respectively. Disease ratings were assessed on September 15 at the beginning maturity (R7) growth stage. White mold disease incidence was assessed by counting the number of plants in each plot with symptoms. For disease severity, each plant was rated according to the following disease category: 0 = no disease, 1 = lateral branches with white mycelium and lesions, 2 = main stem with white mycelium and sclerotia present, and 3 = entire plant wilted/plant death. The disease severity index (DIX) was calculated by DIX = [sum (disease severity score × number of plants)]/[(maximum disease score) × (disease incidence)] × 100. The center rows of each plot were harvested on September 29, and yields were adjusted to 13% moisture. All data were analyzed in SAS 9.4 (SAS Institute, Cary, NC). A generalized linear mixed model analysis of variance was performed using PROC GLIMMIX. Values are least squares means, and values with different letters are significantly different based on a least squares means test (α = 0.05).

In 2022 weather conditions were not favorable for disease, and very little disease developed in plots. White mold was present in the trial but only reached low levels. There were no significant differences between fungicide treatments and the nontreated control for white mold severity on September 15 (Table 38). The fungicide treatment Delaro Complete at R1 followed by Delaro Complete at R3 decreased defoliation over the nontreated control. There was no significant effect of treatment on green stem, harvest moisture, test weight, and yield of soybean.

TABLE 38. *Effect of Fungicide on White Mold, Defoliation, % Green Stem, and Yield of Soybean*

TREATMENT, RATE/ACRE, AND TIMING[z]	WHITE MOLD DIX[y]	DEFOLIATION[x]	GREEN STEM[w] %	HARVEST MOISTURE %	TEST WEIGHT LB/BU	YIELD[v]
Nontreated control/inoculated	0.3	64.3 ab	0.3	14.0	56.5	46.2
Delaro Complete 458 SC 8.0 fl oz at R1	0.0	72.5 a	0.0	13.8	56.6	49.6
Delaro Complete 458 SC 8.0 fl oz at R1 fb Delaro Complete 458 SC 8.0 fl oz at R3	0.0	30.0 c	0.5	14.3	56.0	51.1
Miravis Neo 2.5 SE 13.7 fl oz at R1	0.2	62.5 ab	0.0	14.2	55.2	48.0
Endura 70 WDG 6.0 fl oz at R1	0.0	77.5 a	0.3	13.4	56.6	50.7
Revytek 3.33 LC 8.0 fl oz at R1	0.2	52.5 b	0.8	14.2	55.4	51.4
Omega 12.0 fl oz at R1	0.0	73.8 a	0.5	13.6	56.9	48.9
P-value[u]	0.6078	0.0017	0.5225	0.2630	0.3392	0.2508

[z] Fungicides were applied on July 29 at the beginning pod (R3) growth stage and August 12 at the beginning seed (R5) growth stage. All fungicide treatments contained a nonionic surfactant at a rate of 0.24% v/v. All plots inoculated with *S. sclerotiorum*. fb = followed by.

[y] The disease severity index (DIX) was calculated as DIX = [sum (disease severity score × number of plants)]/[(maximum disease score) × (disease incidence)] × 100 and rated on September 15.

[x] Defoliation was assessed as a percentage (0–100%) of leaf loss in plot and rated on September 15.

[w] Green stem was assessed as a percentage (0–100%) of stems remaining green in plot on September 29.

[v] Yields were adjusted to 13% moisture after harvest on September 29.

[u] All data were analyzed in SAS 9.4 (SAS Institute, Cary, NC). A generalized linear mixed model analysis of variance was performed using PROC GLIMMIX. Values are least squares means, and values with different letters are significantly different based on a least squares means test (α = 0.05).

EVALUATION OF FUNGICIDE PROGRAMS FOR WHITE MOLD IN SOYBEAN IN NORTHWESTERN INDIANA, 2022 (SOY22-23.PPAC)

C. R. Da Silva, S. Shim, and D. E. P. Telenko, Department of Botany and Plant Pathology, Purdue University West Lafayette, IN 47907-2054

SOYBEAN (*GLYCINE MAX* P29A19E)

White mold, *Sclerotinia sclerotiorum*

A trial was established at the Pinney Purdue Agricultural Center (PPAC) in Porter County, Indiana. The experiment was a randomized complete block design with four replications. Plots were 10 feet wide and 30 feet long and consisted of four rows, and the two center rows were used for evaluation. The previous crop was soybean. Standard practices for soybean production in Indiana were followed. Soybean cultivar P29A19E was planted in 30-inch row spacing at a rate of 140,000/acre on May 17. Inoculum of *S. sclerotiorum* was applied on the seedbed at 1.25 g/foot at planting. The field was overhead irrigated weekly at 1 inch unless weekly rainfall was 1 inch or higher to encourage disease. All fungicide applications were applied at 15 gal/acre and 40 psi using a Lee self-propelled sprayer equipped with a 10-foot boom, fitted with six TJ-VS 8002 nozzles spaced 20 inches apart. Fungicides were applied on July 29 at the beginning pod (R3) growth stage and on August 12 at the beginning seed (R5) growth stage. Disease ratings were assessed on September 15 at the beginning maturity (R7) growth stage. White mold disease incidence was assessed by counting the number of plants in each plot with symptoms. For disease severity, each observed plant was rated according to the following disease category: 0 = no disease, 1 = lateral branches with white mycelium and lesions, 2 = main stem with white mycelium and sclerotia present, 3 = entire plant wilted/plant death. The disease severity index (DIX) was calculated as DIX = [sum (disease severity score × number of plants)]/[(maximum disease score) × (disease incidence)] × 100. The center rows of each plot were harvested on September 29, and yields were adjusted to 13% moisture. All data were analyzed in SAS 9.4 (SAS Institute, Cary, NC). A generalized linear mixed model analysis of variance was performed using PROC GLIMMIX. Values are least squares means, and values with different letters are significantly different based on a least squares means test ($\alpha = 0.05$).

In 2022 weather conditions were not favorable for disease development, and very little disease developed in plots. White mold was present in the trial but only reached low levels. There were no significant differences between fungicide treatments and nontreated control for white mold DIX and defoliation on September 15 (Table 39). There was no significant effect of treatment on green stem, harvest moisture, test weight, and yield of soybean.

TABLE 39. *Effect of Fungicide on White Mold, Defoliation, % Green Stem, and Soybean Yield*

TREATMENT, RATE/ACRE, AND TIMING[z]	WHITE MOLD DIX[y]	DEFOLIATION[x]	GREEN STEM[w] %	HARVEST MOISTURE %	TEST WEIGHT LB/BU	YIELD[v] BU/ ACRE
Nontreated control	3.3	47.5	0.6	14.3	56.0	51.9
CX-9032 1.0 qt at R3 fb CX-9032 1.0 qt at R5	0.5	27.5	0.8	14.0	55.9	50.2
Serenade ASO 1.5 qt at R3 fb Serenade ASO 1.5 qt at R5	1.1	42.5	1.7	14.5	56.1	49.4
CX-10250 1.0 fl oz at R3 fb CX-10250 1.0 fl oz at R5	15.4	16.3	0.7	14.3	55.5	49.7
Endura 70 WDG 8.0 fl oz at R3 fb Endura 70 WDG 8.0 fl oz at R5	0.0	27.5	0.5	14.6	55.9	49.8
P-value[u]	0.4844	0.2297	0.5257	0.9122	0.8247	0.4333

[z] Fungicides were applied on July 29 at the beginning pod (R3) growth stage and on August 12 at the beginning seed (R5) growth stage. All fungicide treatments contained a nonionic surfactant (Preference) at a rate of 0.24% v/v. All plots inoculated with *S. sclerotiorum*. fb = followed by.

[y] The disease severity index (DIX) was calculated by multiplying the average number of plants in each severity category by the incidence: DIX = [sum (disease severity score × number of plants)]/[(maximum disease score) × (disease incidence)] × 100 on September 15.

[x] Defoliation was assessed as a percentage (0–100%) of leaf loss in plot on September 15.

[w] Green stem was assessed as a percentage (0–100%) of stem green in plot on September 29.

[v] Yields were adjusted to 13% moisture after harvest on September 29.

[u] All data were analyzed in SAS 9.4 (SAS Institute, Cary, NC). A generalized linear mixed model analysis of variance was performed using PROC GLIMMIX. Values are least squares means, and values with different letters are significantly different based on a least squares means test (α = 0.05).

EVALUATION OF FUNGICIDES FOR WHITE MOLD IN SOYBEAN IN NORTHWESTERN INDIANA, 2022 (SOY22-26.PPAC)

C. R. Da Silva, S. Shim, and D. E. P. Telenko, Department of Botany and Plant Pathology, Purdue University West Lafayette, IN 47907-2054

SOYBEAN (*GLYCINE MAX* P29A19E)

White mold, *Sclerotinia sclerotiorum*

A trial was established at the Pinney Purdue Agricultural Center (PPAC) in Porter County, Indiana. The experiment was a randomized complete block design with four replications. Plots were 10 feet wide and 30 feet long and consisted of four rows, and the two center rows were used for evaluation. The previous crop was soybean. Standard practices for soybean production in Indiana were followed. Soybean cultivar P29A19E was planted in 30-inch row spacing at a rate of 140,000/acre on May 17. Inoculum of *S. sclerotiorum* was applied on the seedbed at 1.25 g/foot at planting. The field was overhead irrigated weekly at 1 inch unless weekly rainfall was 1 inch or higher to encourage disease. Pre-Emergence (PRE-E) applications made with a CO_2 backpack sprayer on May 23. All foliar applications were applied at 15 gal/acre and 40 psi using a Lee self-propelled sprayer equipped with a 10-foot boom, fitted with six TJ-VS 8002 nozzles spaced 20 inches apart. Treatments were applied on June 23, July 14, July 29, and August 23 at V4, beginning bloom (R1), beginning pod (R3), and beginning seed (R5) growth stages, respectively. Disease ratings were assessed on September 15 at beginning maturity (R7) growth stage. White mold disease incidence assessed by counting the number of plants in each plot with symptoms. For disease severity, each plant was rated according to the following disease category: 0 = no disease, 1 = lateral branches with white mycelium and lesions, 2 = main stem with white mycelium and sclerotia present, and 3 = entire plant wilted/plant death. The disease severity index (DIX) was calculated as DIX = [sum (disease severity score X number of plants)]/[(maximum disease score) × (disease incidence)] × 100. The center rows of each plot were harvested on September 29, and yields were adjusted to 13% moisture. All data were analyzed in SAS 9.4 (SAS Institute, Cary, NC). A generalized linear mixed model analysis of variance was performed using PROC GLIMMIX. Values are least squares means, and values with different letters are significantly different based on a least squares means test ($\alpha = 0.05$).

In 2022 weather conditions were unfavorable for disease development, and very little disease developed in plots. White mold was present in the trial but only reached low levels. There were no significant differences between fungicide treatments and the nontreated control for white mold incidence or DIX on September 15 (Table 40). The treatment ORO-079B 2.0 pt PRE-E increased harvest moisture over the nontreated control. There was no significant effect of treatment on test weight and yield of soybean.

TABLE 40. *Effect of Fungicide on White Mold Incidence and Yield of Soybean*

TREATMENT, RATE/ACRE, AND TIMING[z]	WHITE MOLD DI[y] %	WHITE MOLD DIX[y]	HARVEST MOISTURE %	TEST WEIGHT LB/BU	YIELD[x] BU/ACRE
Nontreated control	4.6	111.7	13.6 c	58.2	47.6
ORO-070B 2.0 pt at PRE-E fb ORO-009E 0.4% v/v at R1	5.6	63.4	13.7 c	58.5	47.2
ORO-079B 2.0 pt at PRE-E fb ORO-009E 0.4% v/v at V4 fb ORO-009E 0.4% v/v at R1	9.3	285.0	13.5 c	58.4	46.5
ORO-079B 2.0 pt at PRE-E fb Endura 70 WDG 8.0 fl oz at R1	3.7	43.4	13.9 bc	58.3	52.0
ORO-009E 0.4% v/v at V4 fb ORO-009E 0.4% v/v at R1	7.2	199.1	13.6 c	58.5	45.8
ORO-079B 2.0 pt PRE-E	7.9	229.0	14.7 a	58.6	49.8
Endura 70 WDG 8.0 fl oz at R1	4.8	87.0	13.7 c	58.8	48.1
Endura 70 WDG 8.0 fl oz at R1 + ORO-009E 0.4% v/v at R1	2.5	33.0	13.8 c	58.4	47.4
Endura 70 WDG 8.0 fl oz at R3 + oRO-009E 0.4% v/v at R3	2.9	53.6	13.8 c	58.8	50.7
ORO-369-A 2.0 pt PRE-E fb ORO-009E 0.4% v/v at R1	5.6	209.5	13.6 c	58.5	46.2
ORO-009E 0.4% v/v at R5	7.1	173.3	14.6 ab	58.7	50.2
P-value[w]	0.2446	0.3915	0.0176	0.8577	0.5931

[z] Preemergence (PRE-E) applications were applied on May 23, and foliar applications were applied on June 23, July 14, July 29, and August 23 at V4, beginning bloom (R1), beginning pod (R3), and beginning seed (R5) growth stages, respectively. All plots inoculated with *S. sclerotiorum*. fb = followed by.

[y] White mold disease severity index (DIX) was calculated by the formula DIX = [sum (disease severity score X number of plants)] /[(maximum disease score) X (disease incidence)] × 100 on September 15.

[x] Yields were adjusted to 13% moisture after harvest on September 29.

[v] All data were analyzed in SAS 9.4 (SAS Institute, Cary, NC). A generalized linear mixed model analysis of variance was performed using PROC GLIMMIX. Values are least squares means, and values with different letters are significantly different based on a least squares means test (α = 0.05).

SOUTHWEST PURDUE AGRICULTURAL CENTER (SWPAC)

EVALUATION OF FUNGICIDES FOR FOLIAR DISEASES ON SOYBEAN IN SOUTHWESTERN INDIANA, 2022 (SOY22-02.SWPAC)

S. Shim and D. E. P. Telenko, Department of Botany and Plant Pathology, Purdue University
West Lafayette, IN 47907-2054

SOYBEAN (*GLYCINE MAX* P29A19E)

Frogeye leaf spot, *Cercospora sojina*
Septoria brown spot, *Septoria glycines*

A trial was established at the Southwest Purdue Agricultural Center (SWPAC) in Knox County, Indiana. The experiment was a randomized complete block design with four replications. Plots were 10 feet wide and 30 feet long and consisted of four rows, and the two center rows were used for evaluation. The previous crop was corn. Standard practices for soybean production in Indiana were followed. Soybean cultivar P29A19E was planted in 30-inch row spacing at a rate of 150,000 seed/acre on May 18. All fungicide were applied at 15 gal/acre and 40 psi using a Lee self-propelled sprayer equipped with a 10-foot boom, fitted with six TJ-VS 8002 nozzles spaced 20 inches apart. Fungicides were applied on August 16 at beginning seed (R5) growth stage. Frogeye leaf spot (FLS) and Septoria brown spot (SBS) were rated for disease severity by visually assessing the percentage of symptomatic leaf area in the upper and lower canopies, respectively, on September 12. Canopy greenness was visually assessed as a percentage (0–100%) of canopy green on September 12. The two center rows of each plot were harvested on October 11, and yields were adjusted to 13% moisture. All disease and yield data were analyzed using a mixed model analysis of variance, and means were separated using least squares means test (α = 0.05).

In 2022 weather conditions were not favorable for disease development, and very little disease developed in plots. FLS and SBS were present in the trial but only reached low levels. There was no significant effect of treatment on FLS severity in the upper canopy (Table 41). All fungicides reduced SBS severity over the non-treated control in both the upper and lower canopies on September 12 except Quadris. There was no significant effect of treatment on canopy greenness, harvest moisture, test weight, and yield of soybean.

TABLE 41. *Effect of Treatment on Foliar Diseases, Canopy Greenness, and Yield of Soybean*

TREATMENT, RATE/ACRE, AND TIMING[z]	FLS[y] %	SBS UPPER CANOPY[y] %	SBS LOWER CANOPY[y] %	CANOPY GREEN[x] %	HARVEST MOISTURE %	TEST WEIGHT LB/BU	YIELD[w] BU/ ACRE
Nontreated control	1.8	10.0 a	13.8 a	75.0	8.7	56.2	62.3
Topguard EQ 4.29 5.0 fl oz at R5	1.0	4.5 b	7.0 bc	73.8	8.6	56.1	62.3
Lucento 4.17 SC 5.0 fl oz at R5	0.8	1.3 b	4.3 c	86.3	8.5	55.9	63.0
Trivapro 2.21 SE 13.7 fl oz at R5	1.4	4.0 b	5.5 bc	78.8	8.6	56.1	66.6
Quadris 6.0 fl oz at R5	2.3	7.5 a	12.5 ab	75.0	8.6	56.0	63.0
Veltyma 3.34 S 7.0 fl oz at R5	2.1	2.5 b	5.5 c	81.3	8.6	56.0	67.0
Revytek 3.33 LC 8.0 fl oz at R5	1.8	3.3 b	5.5 c	80.0	9.1	55.7	67.1
Echo 720 36.0 fl oz + Folicur 3.6 F 4.0 fl oz + Topsin 4.5 FL 20.0 fl oz at R5	1.0	2.8 b	6.3 c	75.0	8.6	56.1	62.5
Delaro Complete 458 SC 8.0 fl oz at R5	2.3	2.0 b	5.0 c	77.5	8.6	56.4	63.3
Miravis Neo 2.4 SE 13.7 fl oz at R5	1.5	3.5 b	6.3 c	72.5	8.6	55.7	64.7
Topsin 4.5 FL 20.0 fl oz at R5	1.8	3.5 b	6.3 c	76.3	8.7	55.8	60.6
P-value[v]	0.4557	0.0010	0.0001	0.2688	0.3783	0.9842	0.7552

[z] Fungicides were applied on August 16 at the beginning seed (R5) growth stage. All treatments contained a nonionic surfactant at a rate of 0.25% v/v.

[y] Foliar disease incidence was rated on a scale of 0–100% of plants within a plot with disease symptoms and rated on September 12. FLS = frogeye leaf spot, SBS = Septoria brown spot.

[x] Canopy greenness was visually assessed as a percentage (0–100%) of canopy green on September 12.

[w] Yields were adjusted to 13% moisture after harvest on October 11.

[v] All data were analyzed in SAS 9.4 (SAS Institute, Cary, NC). A generalized linear mixed model analysis of variance was performed using PROC GLIMMIX. Values are least squares means, and values with different letters are significantly different based on a least squares means test (α = 0.05).

EVALUATION OF FUNGICIDES FOR FOLIAR DISEASES ON SOYBEAN IN SOUTHWESTERN INDIANA, 2022 (SOY22-29.SWPAC)

S. Shim and D. E. P. Telenko, Department of Botany and Plant Pathology, Purdue University West Lafayette, IN 47907-2054

SOYBEAN (*GLYCINE MAX* P29A19E)

Frogeye leaf spot, *Cercospora sojina*
Septoria brown spot, *Septoria glycines*

A trial was established at the Southwest Purdue Agricultural Center (SWPAC) in Knox County, IN. The experiment was a randomized complete block design with four replications. Plots were 10 feet wide and 30 feet long and consisted of four rows, and the two center rows were used for evaluation. The previous crop was corn. Standard practices for soybean production in Indiana were followed. Soybean cultivar P29A19E was planted in 30-inch row spacing at a rate of 140,000 seed/acre on May 18. All fungicides were applied at 15 gal/acre and 40 psi using a Lee self-propelled sprayer equipped with a 10-foot boom, fitted with six TJ-VS 8002 nozzles spaced 20 inches apart. Fungicides were applied on August 16 at beginning seed (R5) growth stage. Foliar disease incidence was rated on a scale of 0–100% of plants within a plot with disease symptoms on September 12. Canopy greenness was visually assessed as a percentage (0–100%) of crop canopy on September 12. The two center rows of each plot were harvested on October 11, and yields were adjusted to 13% moisture. All data were analyzed in SAS 9.4 (SAS Institute, Cary, NC). A generalized linear mixed model analysis of variance was performed using PROC GLIMMIX. Values are least squares means, and values with different letters are significantly different based on a least squares means test (α = 0.05).

In 2022 weather conditions were not favorable for disease development, and very little disease developed in plots. Frogeye leaf spot (FLS) and Septoria brown spot (SBS) were present in the trial but only reached low levels. All fungicides significantly reduced SBS severity in the lower canopy compared to the nontreated control on September 12 (Table 42). There was no significant difference between treatments and FLS severity and SBS in the upper canopy on September 12. All fungicides significantly increased canopy greenness compared to the nontreated control. There was no significant effect of treatment on harvest moisture, test weight, and yield of soybean.

TABLE 42. *Effect of Fungicide on Foliar Disease, Canopy Greenness, and Yield of Soybean*

TREATMENT AND RATE/ACRE[z]	FLS[Y] %	SBS UPPER CANOPY[Y] %	SBS LOWER CANOPY[Y] %	CANOPY GREEN[x] %	HARVEST MOISTURE %	TEST WEIGHT LB/BU	YIELD[w] BU/ACRE
Nontreated control	0.7	2.4	10.4 a	67.5 c	8.8	57.6	62.2
Delaro Complete 458 SC 8.0 fl oz	0.3	1.0	3.3 b	85.0 a	8.8	56.8	57.1
Lucento 4.17 SC 5.0 fl oz	0.2	0.9	4.5 b	72.5 bc	8.7	56.2	58.7
Trivapro 2.21 SE 13.7 fl oz	0.3	1.3	4.5 b	73.8 bc	8.8	56.3	55.0
Miravis Neo 2.5 SE 13.7 fl oz	0.7	0.9	3.3 b	78.8 ab	8.9	55.8	56.2
Revytek 3.33 LC 8.0 fl oz	0.8	1.0	3.0 b	75.0 bc	9.1	55.3	60.7
P-value[v]	0.0718	0.0596	0.0004	0.0075	0.6780	0.3678	0.2437

[z] Fungicides were applied on August 16 at the beginning seed (R5) growth stage, and all treatments contained a nonionic surfactant (Preference) at a rate of 0.25%.

[y] Foliar disease incidence was rated on a scale of 0–100% of plants within a plot with disease symptoms on September 12. FLS = frogeye leaf spot, SBS = Septoria brown spot.

[x] Canopy greenness was visually assessed as a percentage (0–100%) of crop canopy green on September 12.

[w] Yields were adjusted to 13% moisture after harvest on October 11.

[v] All data were analyzed in SAS 9.4 (SAS Institute, Cary, NC). A generalized linear mixed model analysis of variance was performed using PROC GLIMMIX. Values are least squares means, and values with different letters are significantly different based on a least squares means test ($\alpha = 0.05$).

EVALUATION OF FOLIAR FUNGICIDES FOR SCAB MANAGEMENT IN SOUTHERN INDIANA, 2022 (WHT22-04.SWPAC)

M. S. Mizuno, S. Shim, and D. E. P. Telenko, Department of Botany and Plant Pathology, Purdue University West Lafayette, IN 47907-2054

WHEAT (*TRITICUM AESTIVUM* P25R40)

Fusarium head blight, *Fusarium graminearum*

A trial was established at the Southwest Purdue Agricultural Center (SWPAC) in Knox County, Indiana. The experiment was a randomized complete block design with four replications. Plots were 7.5 feet wide and 20 feet long and consisted of 12 rows spaced 7.5 inches apart, and the center of each plot was used for evaluation. The previous crop was soybean. On November 4, 2021, wheat cultivar P25R40 was drilled at 7.5-inch spacing. Harmony Extra at 0.8 oz/acre plus AMS at 2 lb/acre plus NIS at 0.25% v/v was applied on March 29, 2022, for weed management. All fungicide applications were applied at 15 gal/acre and 40 psi using a CO_2 backpack sprayer equipped with a 10-foot boom, fitted with six TJ-VS 8002 nozzles spaced 20-inch apart and directed forward and backward at a 45-degree angle at 3.0 mph. Fungicides were applied on May 11 and May 17 at Feekes growth stages 10.5.1 and 10.5.1 + 5 days after 10.5.1, respectively. All plots were inoculated with a mixture of isolates of *Fusarium graminearum* endemic to Indiana on May 11. The spore suspension (50,000 spores/ml) was applied at 300 ml/plot with the CO_2 backpack sprayer. Disease ratings were assessed on May 31. Fusarium head blight (FHB) incidence was measured as the number of infected heads out of 60 plants in each plot and calculated as a percentage. FHB severity was rated by visually assessing the percentage of the infected head. The FHB index was calculated as (% FHB incidence multiplied by average FHB severity)/100 per plot. The eight center rows of each plot were harvested on June 21, and yields were adjusted to 13.5% moisture. Data were subjected to mixed model analysis of variance (SAS 9.4, 2019), and means were compared using a least squares means test (α = 0.05).

In 2022, weather conditions were moderately favorable for FHB. FHB incidence (DI), severity (DS), and index were significantly reduced by all fungicides (Table 43). The concentration of deoxynivalenol (DON) was reduced over the nontreated control for all treatments. There was no difference in FHB incidence, percentage of fusarium damaged kernels (FDK), and yield of wheat.

TABLE 43. *Effect of Fungicide on Fusarium Head Blight, Deoxynivalenol (DON), Fusarium Damaged Kernels (FDK), and Yield of Wheat*

TREATMENT AND RATE/ACRE[z]	FHB DI[y]	FHB DS[x]	FHB INDEX[w]	FDK[v] %	DON[u] PPM	YIELD[t] BU/ ACRE
Nontreated control	85.0 a	3.1 a	2.6 a	7.0	0.8 a	109.1
Prosaro 421 SC 6.5 fl oz at 10.5.1	53.8 b	2.0 bcd	1.1 bc	7.0	0.4 b	122.3
Caramba 90 EC 13.5 fl oz at 10.5.1	60.0 b	2.2 bc	1.3 b	5.0	0.3 bc	110.0
Miravis Ace 5.2 SC 13.7 fl oz at 10.5.1	35.4 c	2.0 bcd	0.8 cd	5.5	0.3 bcd	116.1
Prosaro Pro 10.3 fl oz at 10.5.1	52.9 b	2.3 b	1.2 bc	6.0	0.3 bc	114.3
Sphaerex (BAS 84000F) 7.3 fl oz at 10.5.1	52.9 b	2.0 bcd	1.1 bc	5.8	0.3 bcd	111.6
Miravis Ace 5.2 SC 13.7 fl oz at 10.5.1 fb Prosaro 421 SC 6.5 fl oz at 10.5.1 + 5 d	24.2 c	1.4 d	0.4 d	4.3	0.2 cd	109.1
Miravis Ace 5.2 SC 13.7 fl oz at 10.5.1 fb Sphaerex 7.3 fl oz 10.5.1 + 5 d	23.8 c	1.6 cd	0.4 d	3.5	0.1 d	119.1
Miravis Ace 5.2 SC 13.7 fl oz at 10.5.1 fb Tebuconazole 4.0 fl oz at 10.51 + 5 d	21.7 c	1.7 bcd	0.4 d	5.3	0.2 cd	118.1
P-value[s]	0.0001	0.0043	0.0001	0.0515	0.0001	0.2957

[z] Fungicide treatments were applied on May 11 and May 17 at Feekes growth stages 10.5.1 and 10.5.1 + 5 d, respectively. All treatments contained a nonionic surfactant (Preference) at a rate of 0.125% v/v. All plots were inoculated with *Fusarium graminearum* spore suspension (50,000 spores/ml) after the treatment at Feekes 10.5.1. Spore suspension were applied at 300 ml/plot with handheld sprayer. fb = followed by.

[y] FHB disease incidence (DI) was measured as the number of infected heads out of 60 plants in each plot and calculated as a percentage on May 31.

[x] FHB disease severity (DS) was rated by visually assessing the percentage of the infected head on May 31. FHB = Fusarium head blight.

[w] The FHB index was calculated as (FHB incidence multiplied by average FHB severity)/100 per plot on May 31.

[v] Fusarium damaged kernels (FDK) were visually assessed as a percentage of Fusarium damaged heads.

[u] Analysis of mycotoxin deoxynivalenol (DON) was completed by the University of Minnesota DON Testing Lab.

[t] Yields were adjusted to 13.5% moisture after harvest on June 21.

[s] All data were analyzed in SAS 9.4 (SAS Institute, Cary, NC). A generalized linear mixed model analysis of variance was performed using PROC GLIMMIX. Values are least squares means, and values with different letters are significantly different based on a least squares means test (α = 0.05).

EVALUATION OF FOLIAR FUNGICIDES AND CULTIVARS FOR SCAB MANAGEMENT IN SOUTHERN INDIANA, 2022 (WHT22-05.SWPAC)

K. M. Goodnight, S. Shim, and D. E. P. Telenko, Department of Botany and Plant Pathology, Purdue University West Lafayette, IN 47907-2054

WHEAT (*TRITICUM AESTIVUM* P25R40 AND P25R61)

Fusarium head blight, *Fusarium graminearum*

A trial was established at the Southwest Purdue Agricultural Center (SWPAC) in Knox County, Indiana. The experiment was a strip-plot design with four replications. Plots were 7.5 feet wide and 20 feet long and consisted of 12 rows spaced 7.5 inches apart, and the center of each plot was used for evaluation. The previous crop was corn. Wheat cultivars P25R40 and P25R61 were planted at 7.5-inch spacing using a drill on November 4, 2021. Harmony Extra at 0.8 oz/acre plus AMS at 2 lb/acre plus NIS at 0.25% v/v was applied on March 29, 2020, for weed management. All fungicide applications were applied at 15 gal/acre and 40 psi using a CO_2 backpack sprayer equipped with a 10-foot boom, fitted with six TJ-VS 8002 nozzles spaced 20 inches apart and directed forward and backward at a 45-degree angle at 4.0 mph. Fungicides were applied on May 11 at Feekes growth stage 10.5.1. A mixture of isolates of *Fusarium graminearum* endemic to Indiana were used to inoculate plots on May 11. The spore suspension (50,000 spores/ml) was applied at 300 ml/plot with the CO_2 handheld sprayer. Disease ratings were assessed on May 31. Fusarium head blight (FHB) incidence was measured as the number of infected heads out of 60 plants in each plot and calculated as a percentage. FHB severity was rated by visually assessing the percentage of the infected head. The FHB index was calculated as (% FHB incidence multiplied by average FHB severity)/100 per plot. Disease severity on leaves was rated by visually assessing the percentage of symptomatic leaf tissue on five flag leaves per plot for leaf blotch. Values for each plot were averaged before analysis. The eight center rows of each plot were harvested with a Kincaid 8XP combine on June 21, and yields were adjusted to 13.5% moisture. Data were subjected to mixed model analysis of variance (SAS 9.4, 2019), and means were compared using least squares means test ($\alpha = 0.05$).

In 2022, weather conditions were moderately favorable for FHB. FHB was the most prominent disease, and there was little to no leaf blotch detected. The scab resistant cultivar P25R61 had significantly less FHB, deoxynivalenol (DON), test weight, harvest moisture, and yield as compared to the susceptible P25R40 cultivar (Table 44). FHB incidence, severity, and index were reduced by Miravis Ace at 10.5.1 and Prosaro Pro at 10.5.1 as compared to the nontreated, inoculated control. No significant differences were detected between treatments and nontreated controls for DON, FDK, and wheat yield.

TABLE 44. *Effect of Cultivar and Fungicide on Fusarium Head Blight (FHB), Deoxynivalenol (DON), Fusarium Damaged Kernels (FDK), and Yield of Wheat*

CULTIVAR, TREATMENT, AND RATE/ACRE[z]	FHB DI[y]	FHB DS[x]	FHB INDEX[w]	DON[v] PPM	FDK[u] %	YIELD[t] BU/ ACRE
Cultivar						
P25R40 (scab susceptible)	62.9 a[s]	2.6 a	1.7 a	0.541 a	8.0	95.0 a
P25R61 (scab resistant)	31.5 b	1.7 b	0.6 b	0.070 b	7.8	86.7 b
Fungicide and rate/Acre						
Nontreated control, inoculated control	58.1 a	2.6 a	1.6 a	0.363	8.0	91.1
Nontreated, noninoculated control	55.6 ab	2.4 ab	1.4 ab	0.339	8.4	92.1
Prosaro 421 SC 6.5 fl oz at 10.5.1	43.1 bc	2.3 ab	1.1 bcd	0.324	7.3	93.0
Miravis Ace 5.2 SC 13.7 fl oz at 10.5.1	30.8 c	1.7 c	0.6 d	0.223	7.3	88.5
Prosaro Pro 10.3 fl oz at 10.5.1	41.3 c	1.9 bc	0.8 d	0.249	8.0	87.7
Sphaerex 7.3 fl oz at 10.5.1	54.2 ab	2.2 abc	1.3 abc	0.338	8.8	93.0
P-value cultivar[s]	0.0001	0.0001	0.0001	0.0001	0.5872	0.0033
P-value fungicide	0.0005	0.0484	0.0023	0.4740	0.1703	0.7612
P-value cultivar*fungicide	0.9163	0.9968	0.6664	0.3674	0.7931	0.0240

[z] Fungicide treatments were applied on May 11 at Feekes growth stage 10.5.1. All treatments contained a nonionic surfactant at a rate of 0.125% v/v. All plots were inoculated with *Fusarium graminearum* spore suspension (50,000 spores/ml) after the treatment at Feekes 10.5.1. Spore suspension applied at 300 ml/plot with handheld sprayer on May 11.

[y] FHB disease incidence (DI) was measured as the number of infected heads out of 60 plants in each plot and calculated as a percentage on May 31.

[x] FHB disease severity (DS) was rated by visually assessing the percentage of the infected head on May 31. FHB = Fusarium head blight.

[w] FHB index was calculated as (FHB incidence multiplied by average FHB severity)/100 per plot on May 31.

[v] Analysis of mycotoxin deoxynivalenol (DON) was completed by the University of Minnesota DON Testing Lab.

[u] Fusarium damaged kernels (FDK) were visually assessed as a percentage of Fusarium damaged heads.

[t] Yields were adjusted to 13.5% moisture after harvest on June 21.

[s] All data were analyzed in SAS 9.4 (SAS Institute, Cary, NC). A generalized linear mixed model analysis of variance was performed using PROC GLIMMIX. Values are least squares means, and values with different letters are significantly different based on a least squares means test (α = 0.05).

DAVIS PURDUE AGRICULTURAL CENTER (DPAC)

FIELD-SCALE EVALUATION OF FUNGICIDES FOR FOLIAR DISEASE IN CORN IN CENTRAL INDIANA, 2022 (COR22-08.DPAC)

K. G. Waibel, S. C. Boyer, and D. E. P. Telenko, Department of Botany and Plant Pathology, Purdue University West Lafayette, IN 47907-2054

CORN (*ZEA MAYS* P0574AM)

Gray leaf spot, *Cercospora zeae-maydis*

A trial was established at the Davis Purdue Agricultural Center (DPAC) in Randolph County, Indiana. The experiment was a randomized complete block design with four replications. Plots were 30 feet wide and 500 feet long and consisted of 12 rows, and the two center rows were used for evaluation. The previous crop was soybean. Standard practices for nonirrigated corn production in Indiana were followed. Corn hybrid P0574AMXT was planted in 30-inch row spacing at a rate of 31,000 seeds/acre on May 13. All fungicide applications were applied at 20 gal/acre and 50 psi using either a Raven plot sprayer or a Case IH Patriot sprayer. Fungicides were applied on June 16, June 22, and August 1 at V8, V10, and blister (R2) growth stages, respectively. Gray leaf spot (GLS) was assessed on August 12 at the milk (R3) growth stage and on August 22 at the dent (R5) growth stage. Disease severity was rated by visually assessing the percentage of symptomatic leaf area at the ear leaf. Ten plants in three locations were assessed in each plot and averaged before analysis. Canopy greenness was visually assessed as a percentage (0–100%) of crop canopy green on September 7. The 12 rows of each plot were harvested on October 19, and yields were adjusted to 15.5% moisture. All data were analyzed in SAS 9.4 (SAS Institute, Cary, NC). A generalized linear mixed model analysis of variance was performed using PROC GLIMMIX. Values are least squares means, and values with different letters are significantly different based on a least squares means test ($\alpha = 0.05$).

In 2022, weather conditions were not favorable for disease. GLS was the most prominent disease in the trial and reached low severity. All treatments reduced GLS severity over the nontreated control on August 12 (Table 45). On August 22, GLS severity was significantly lower in plots treated with Delaro at the V8 and V10 growth stages. The V8 application had significantly lower GLS severity than all other treatments on August 12 and August 22. Canopy greenness was significantly higher in the V8 and nontreated plots over the treatments at V10 and R2. There was no significant difference between treatments for harvest moisture and yield of corn.

TABLE 45. *Effect of Fungicide on Gray Leaf Spot (GLS), Canopy Greenness, and Yield of Corn*

TREATMENT, RATE/ACRE, AND TIMING[z]	GLS %[y] AUGUST 12	GLS %[y] AUGUST 22	CANOPY GREEN[x] %	HARVEST MOISTURE %	YIELD[w] BU/ ACRE
Nontreated control	0.4 a	1.0 a	76.0 a	16.3	192.6
Delaro 325 SC 8.0 fl oz at V8	0.0 c	0.2 c	77.0 a	16.2	195.2
Delaro 325 SC 8.0 fl oz at V10	0.2 b	0.7 b	74.2 b	16.2	191.2
Delaro 325 SC 8.0 fl oz at R2	0.2 b	1.0 a	72.5 b	16.2	196.0
P-value[v]	0.0001	0.0002	0.0451	0.5990	0.5158

[z] Fungicides were applied on June 16, June 22, and August 1 at the V8, V10, and blister (R2) growth stages, respectively.

[y] Disease severity was visually assessed as a percentage (0–100%) of symptomatic leaf area on ear leaf. Ten leaves were assessed in three locations per plot and averaged before analysis. GLS = gray leaf spot.

[x] Canopy greenness was visually assessed as a percentage (0–100%) of crop canopy green on September 7.

[w] Yields were adjusted to 15.5% moisture after harvest on October 19.

[v] All data were analyzed in SAS 9.4 (SAS Institute, Cary, NC). A generalized linear mixed model analysis of variance was performed using PROC GLIMMIX. Values are least squares means, and values with different letters are significantly different based on a least squares means test (α = 0.05).

FIELD-SCALE FUNGICIDE TIMING COMPARISON FOR FOLIAR DISEASES ON SOYBEAN IN CENTRAL INDIANA, 2022 (SOY22-07.DPAC)

K. G. Waibel, J. Boyer, and D. E. P. Telenko, Department of Botany and Plant Pathology, Purdue University West Lafayette, IN 47907-2054

SOYBEAN (*GLYCINE MAX* P29A19E)

Frogeye leaf spot, *Cercospora sojina*
Septoria brown spot, *Septoria glycines*
Downy mildew, *Peronospora manshurica*

A trial was established at the Davis Purdue Agricultural Center (DPAC) in Randolph County, Indiana. The experiment was a randomized complete block design with four replications. Plots were 30 feet wide and 485 feet long and consisted of 24 rows, and the two center rows were used for evaluation. The previous crop was corn. Standard practices for nonirrigated soybean production in Indiana were followed. Soybean cultivar P29A19E was planted in 15-inch row spacing at a rate of 150,000 seeds/acre on May 23. All fungicide applications were applied at 20 gal/acre and 50 psi using a Raven plot sprayer or Case IH 2240 sprayer. Fungicides were applied on June 30, July 25, and August 1 at the V4, beginning pod (R3), and beginning seed (R5) growth stages, respectively. Disease ratings were assessed on August 22 at the beginning seed (late R5) and full seed (early R6) growth stages. Septoria brown spot (SBS), frogeye leaf spot (FLS), and downy mildew (DM) were rated for disease severity by visually assessing the percentage of symptomatic leaf area in the upper and lower canopies. The soybeans were harvested on October 6, and yields were adjusted to 13% moisture. All data were averaged before analysis in SAS 9.4 (SAS Institute, Cary, NC). All disease and yield data were analyzed using a mixed model analysis of variance, and means were separated using a least squares means test (α = 0.05).

In 2022 weather conditions were not favorable for disease development, and very little disease developed in plots. DM was the most prominent disease and reached low severity. There was no significant difference between treatments and nontreated controls for FLS, DM, SBS, harvest moisture, and yield of soybean (Table 46).

TABLE 46. *Effect of Fungicide Timing on Foliar Disease Severity and Yield of Soybean*

TREATMENT, RATE/ACRE, AND TIMING[z]	FLS[y] UPPER CANOPY %	FLS[y] LOWER CANOPY %	DM[y] UPPER CANOPY %	SBS[y] LOWER CANOPY %	HARVEST MOISTURE %	YIELD[x] BU/ ACRE
Nontreated control	0.4	0.3	6.8	2.8	9.5	62.2
Delaro 325 SC 12 fl oz at V4	0.0	0.1	4.9	2.8	9.5	60.3
Delaro 325 SC 12 fl oz at R3	0.6	0.5	5.8	3.4	9.5	59.3
Delaro 325 SC 12 fl oz at R5	0.1	0.1	5.6	2.7	9.5	61.3
P-value[w]	0.2443	0.5799	0.0833	0.3811	0.9467	0.3118

[z] Fungicides were applied on June 30, July 25, and August 1 at the V4, beginning pod (R3), and beginning seed (R5) growth stages, respectively.

[y] Foliar disease severity was visually rated on a scale of 0–100% of the upper and lower canopies with disease symptoms on August 22. FLS = frogeye leaf spot, DM = downy mildew, SBS = Septoria brown spot.

[x] Yields were adjusted to 13% moisture after harvest on October 6.

[w] All data were analyzed in SAS 9.4 (SAS Institute, Cary, NC). A generalized linear mixed model analysis of variance was performed using PROC GLIMMIX. Values are least squares means, and values with different letters are significantly different based on a least squares means test ($\alpha = 0.05$).

NORTHEAST PURDUE AGRICULTURAL CENTER (NEPAC)

FIELD-SCALE FUNGICIDE TIMING COMPARISON FOR FOLIAR DISEASES IN CORN IN NORTHEASTERN INDIANA, 2022 (COR22-09.NEPAC)

K. G. Waibel, S. C. Boyer, and D. E. P. Telenko, Department of Botany and Plant Pathology, Purdue University West Lafayette, IN 47907-2054

CORN (*ZEA MAYS* P0574AM)

Gray leaf spot, *Cercospora zeae-maydis*

A trial was established at the Northeast Purdue Agricultural Center (NEPAC) in Whitley County, Indiana. The experiment was a randomized complete block design with four replications. Plots were 30 feet wide and 360 feet long and consisted of 12 rows, and the two center rows were used for evaluation. The previous crop was soybean. Standard practices for nonirrigated corn production in Indiana were followed. Corn hybrid P0574AM was planted in 30-inch row spacing at a rate of 34,000 seeds/acre on May 20. Fungicide treatments were applied on June 23, July 14, July 29, August 12, and August 18 at the V6, V10, tassel/silk (VT/R1), blister (R2), and milk (R3) growth stages, respectively. Disease ratings were assessed on August 24 at the dent (R5) growth stage. Gray leaf spot (GLS) was rated for disease severity by visually assessing the percentage (0–100%) of symptomatic leaf area on the ear leaf on 10 plants at three locations in each plot. Lodging was assessed by pushing 10 plants from shoulder height at a 45-degree angle at three locations in each plot and recording the number with snapped or bent stalks. Plants lodged from severe wind were totaled out of 10 in each location and added to the lodging total. Canopy greenness was visually assessed as a percentage (0–100%) of canopy green on September 7. The trial was harvested on October 10, and yields were adjusted to 15.5% moisture. Data were averaged before analysis and subjected to mixed modeli analysis of variance (SAS 9.4, 2019). A generalized linear mixed model analysis of variance was performed using PROC GLIMMIX. Values are least squares means, and values with different letters are significantly different based on a least squares means test (α = 0.05).

In 2022, weather conditions were not favorable for disease. GLS reached low severity. Headline Amp applied at V6, VT/R1, and R3 significantly reduced GLS severity over the nontreated control on August 24 (Table 47). No treatment was significantly different from the nontreated control for lodging, but the Headline AMP application at V10 had significantly reduced lodging over applications made at V6 and VT/R1. There was no significant effect of fungicide timing on canopy greenness, harvest moisture, and yield of corn.

TABLE 47. *Effect of Fungicide Timing on Foliar Diseases Severity, Lodging, Canopy Greenness, and Yield of Corn*

TREATMENT, RATE/ACRE, AND TIMING[z]	GLS[y] %	CANOPY GREEN[x] %	LODGING[w] %	HARVEST MOISTURE %	YIELD[v] BU/ ACRE
Nontreated control	0.08 a	79.6	9.8 abc	19.9	242.4
Headline AMP 1.68 SC 10.0 fl oz at V6	0.04 bc	82.9	10.4 ab	19.7	234.7
Headline AMP 1.68 SC 10.0 fl oz at V10	0.08 ab	80.8	4.2 c	19.8	235.6
Headline AMP 1.68 SC 10.0 fl oz at VT/R1	0.00 c	77.1	15.4 a	20.3	240.6
Headline AMP 1.68 SC 10.0 fl oz at R2	0.09 a	80.8	6.7 bc	19.9	242.4
Headline AMP 1.68 SC 10.0 fl oz at R3	0.01 b	82.9	6.5 bc	20.3	243.6
P-value[u]	0.0006	0.1365	0.0135	0.7679	0.4237

[z] Fungicide treatments were applied on June 23, July 14, July 29, August 12, and August 18 at V6, V10, tassel/silk (VT/R1), blister (R2), and milk (R3) growth stages, respectively.

[y] Disease severity was visually assessed as a percentage (0–100%) of symptomatic leaf area on the ear leaf. Ten leaves were assessed per plot and averaged on August 24.

[x] Canopy greenness was visually assessed as a percentage (0–100%) on September 7.

[w] Lodging was assessed as a percentage (0–100%) of lodged stalks present in the plot and lodged stalks when pushed from shoulder height to 45° from vertical on September 7.

[v] Yields were adjusted to 15.5% moisture after harvest on October 10.

[u] All data were analyzed in SAS 9.4 (SAS Institute, Cary, NC). A generalized linear mixed model analysis of variance was performed using PROC GLIMMIX. Values are least squares means, and values with different letters are significantly different based on a least squares means test (α = 0.05).

EVALUATION OF XYWAY 2X2 APPLICATION FOR FOLIAR DISEASES IN CORN IN NORTHEASTERN INDIANA, 2022 (COR22-13.NEPAC)

K. G. Waibel, S. C. Boyer, and D. E. P. Telenko, Department of Botany and Plant Pathology, Purdue University West Lafayette, IN 47907-2054

CORN (*ZEA MAYS*, 5794V2P)

Gray leaf spot, *Cercospora zeae-maydis*

A trial was established at the Northeast Purdue Agricultural Center (NEPAC) in Whitley County, Indiana. The experiment was a randomized complete block design with eight replications. Plots were 30 feet wide and 360 feet long and consisted of 12 rows, and the two center rows were used for evaluation. The previous crop was soybean. Standard practices for nonirrigated corn production in Indiana were followed. Corn hybrid 5794V2P was planted in 30-inch row spacing at a rate of 34,000 seeds/acre on May 20. Xyway 15.2 fl oz/acre was applied with the starter fertilizer in a 2x2 configuration (two inches below and two inches to the side of the seed furrow) with 28% nitrogen and ammonium thiosulfate at 13.4 gal/acre at planting. Disease ratings were assessed on August 24 at the dent (R5) growth stage. Gray leaf spot (GLS) was rated by visually by assessing the percentage (0–100%) of symptomatic leaf area on the ear leaf on 10 plants at three locations in each plot. Canopy greenness was visually assessed as a percentage (0–100%) of canopy green on September 7. Lodging was assessed by pushing 10 plants from shoulder height at a 45-degree angle at three locations in each plot and recording the number with snapped or bent stalks. Plants lodged from severe wind were totaled out of 10 in each location and added to the lodging percentage. The trial was harvested on October 11, and yields were adjusted to 15.5% moisture. All data were analyzed in SAS 9.4 (SAS Institute, Cary, NC). A generalized linear mixed model analysis of variance was performed using PROC GLIMMIX. Values are least squares means, and values with different letters are significantly different based on a least squares means test ($\alpha = 0.05$).

In 2022 weather conditions were not favorable for diseases, and very little disease developed in plots. GLS was the most prominent disease and reached low severity. The 2x2 application of Xyway at planting had significantly lower GLS compared to the nontreated control. (Table 48). There were no significant differences between treatments for canopy greenness and lodging on September 7. Yield was significantly higher in the Xyway-treated plots as compared to the nontreated control.

TABLE 48. *Effect of Fungicide on Foliar Diseases Severity, Canopy Greenness, Lodging, and Yield of Corn*

TREATMENT AND RATE/ACRE[z]	GLS[y] %	CANOPY GREEN[x] %	LODGING[w] %	HARVEST MOISTURE %	YIELD[v] BU/ ACRE
Nontreated control	0.009 a	74.6	18.3	20.9	254.9 b
Xyway LFR 15.2 fl oz applied 2x2	0.001 b	76.0	17.6	20.6	266.5 a
P-value[u]	0.0442	0.3042	0.8054	0.1318	0.0135

[z] Xyway 15.2 fl oz was applied in starter fertilizer in 2x2 with 28% nitrogen and ammonium thiosulfate at 13.4 gal/acre at planting on May 20.

[y] Disease severity was visually assessed as a percentage (0–100%) of symptomatic leaf area on ear leaf on August 24. GLS=gray leaf spot.

[x] Canopy greenness was visually assessed as a percentage (0–100%) of canopy green on September 7.

[w] Lodging was assessed as a percentage (0–100%) of lodged stalks present in the plot and lodged stalks when pushed from shoulder height to 45° from vertical on September 7.

[v] Yields were adjusted to 15.5% moisture after harvest on October 11.

[u] All data were analyzed in SAS 9.4 (SAS Institute, Cary, NC). A generalized linear mixed model analysis of variance was performed using PROC GLIMMIX. Values are least squares means, and values with different letters are significantly different based on a least squares means test (α = 0.05).

FIELD-SCALE FUNGICIDE TIMING FOR FOLIAR DISEASES ON SOYBEAN IN NORTHEASTERN INDIANA, 2022 (SOY22-09.NEPAC)

K. G. Waibel, J. Boyer, and D. E. P. Telenko, Department of Botany and Plant Pathology, Purdue University West Lafayette, IN 47907-2054

SOYBEAN (*GLYCINE MAX* P29A19E)

Frogeye leaf spot, *Cercospora sojina*
Septoria brown spot, *Septoria glycines*
Downy mildew, *Peronospora manshurica*
White Mold, *Sclerotinia sclerotiorum*

A trial was established at the Northeast Purdue Agricultural Center (NEPAC) in Whitley County, Indiana. The experiment was a randomized complete block design with four replications. Plots were 30 feet wide and 390 feet long. The previous crop was corn. Standard practices for nonirrigated soybean production in Indiana were followed. Soybean cultivar P29A19E was drilled in 7.5-inch row spacing at a rate of 175,000 seeds/acre on May 12. Fungicide treatments were applied on June 22, July 10, July 29, and August 12 at the V4, beginning bloom (R1), beginning pod (R3), and beginning pod (R3) growth stages, respectively. Disease ratings were assessed on August 24 at the early full seed (R6) growth stage. Septoria brown spot (SBS), frogeye leaf spot (FLS), and downy mildew (DM) were rated for disease severity by visually assessing the percentage of symptomatic leaf area in the upper and lower canopies in three locations in each plot. White mold was rated by visually assessing the number of infected plants within a 38-foot–diameter radius at three locations in each plot. The soybeans were harvested on October 1, and yields were adjusted to 13% moisture All data were analyzed in SAS 9.4 (SAS Institute, Cary, NC). A generalized linear mixed model analysis of variance was performed using PROC GLIMMIX. Values are least squares means, and values with different letters are significantly different based on a least squares means test (α = 0.05).

In 2022 weather conditions were not favorable for diseases, and very little disease developed in plots. All timings of Miravis Top significantly reduced FLS severity in the upper and lower canopy over the nontreated control (Table 49). All treatments significantly reduced SBS severity compared to the nontreated control. Miravis Top at V4 resulted in significantly lower SBS severity than application at R1 and R3, but the V4 treatment was not significantly different from the R5 application. No differences were detected between treatments for DM. There was no significant difference between treatments for white mold. Miravis Top applied at R3 significantly increased yield over the nontreated control and other fungicide application timings.

TABLE 49. *Effect of Fungicide Timing on Foliar Disease Severity and Yield of Soybean*

TREATMENT, RATE/ACRE, AND TIMING[z]	FLS[Y] UPPER CANOPY %	FLS[Y] LOWER CANOPY %	SBS[Y] %	DM[Y] %	WHITE MOLD[X] # PLANTS	YIELD[W] BU/ACRE
Nontreated control	1.5 a	0.9 a	2.3 a	1.7	0.6	61.0 b
Miravis Top 1.67 SC 13.7 oz at V4	0.6 b	0.1 b	0.2 c	1.7	1.6	58.6 b
Miravis Top 1.67 SC 13.7 oz at R1	0.1 b	0.1 b	1.3 b	0.7	0.2	61.1 b
Miravis Top 1.67 SC 13.7 oz at R3	0.6 b	0.2 b	1.4 b	1.3	0.6	64.9 a
Miravis Top 1.67 SC 13.7 oz at R5	1.7 b	0.2 b	0.8 bc	1.9	2.3	61.2 b
P-value[v]	0.0074	0.0039	0.0013	0.1790	0.2134	0.0253

[z] Fungicide treatments were applied on June 22, July 10, July 29, and August 12 at the V4, beginning bloom (R1), beginning pod (R3), and beginning pod (R3) growth stages, respectively.

[y] Foliar disease severity was visually rated on a scale of 0–100% of the upper and lower canopies with disease symptoms on August 24. FLS = frogeye leaf spot, SBS = Septoria brown spot in lower canopy, DM = downy mildew upper canopy.

[x] White mold was rated by visually assessing the number of infected plants at three locations in each plot and then averaged on August 24.

[w] Yields were adjusted to 13% moisture after harvest on October 1.

[v] All data were analyzed in SAS 9.4 (SAS Institute, Cary, NC). A generalized linear mixed model analysis of variance was performed using PROC GLIMMIX. Values are least squares means, and values with different letters are significantly different based on a least squares means test ($\alpha = 0.05$).

SOUTHEAST PURDUE AGRICULTURAL CENTER (SEPAC)

FIELD-SCALE EVALUATION OF FUNGICIDE TIMING FOR FOLIAR DISEASE IN CORN IN SOUTHEASTERN INDIANA, 2022 (COR22-10.SEPAC)

K. G. Waibel, J. R. Wahlman, A. Helms, and D. E. P. Telenko, Department of Botany and Plant Pathology, Purdue University West Lafayette, IN 47907-2054

CORN (*ZEA MAYS* P1077)

Gray leaf spot, *Cercospora zeae-maydis*

A trial was established at the Southeast Purdue Agricultural Center (SEPAC) in Jennings County, Indiana. The experiment was a randomized complete block design with four replications. Plots were 30 feet wide and 600 feet long and consisted of 12 rows, and the two center rows were used for evaluation. The previous crop was soybean. Standard practices for nonirrigated corn production in Indiana were followed. Corn hybrid P1077 was planted in 30-inches row spacing at a rate of 33,000 seeds/acre on May 12. All fungicide applications were applied at 20 gal/acre and 35 psi using an Apache 720 sprayer. Fungicides were applied on June 15, July 19, and August 4 at the V6, tassel (VT), and milk (R3) growth stages, respectively. Disease ratings were assessed on August 2 at the milk (R3) and on August 16 at the dent (R5) growth stages. Gray leaf spot (GLS) was rated for disease severity by visually assessing the percentage (0–100%) of symptomatic leaf area on the ear leaf. Ten plants in three locations were assessed in each plot and averaged before analysis. Canopy greenness was visually assessed as a percentage (0–100%) of canopy green on September 2. The 12 rows of each plot were harvest on October 5, and yields were adjusted to 15.5% moisture. Data were subjected to mixed modeli analysis of variance (SAS 9.4, 2019). A generalized linear mixed model analysis of variance was performed using PROC GLIMMIX. Values are least squares means, and values with different letters are significantly different based on a least squares means test (α = 0.05).

In 2022 weather conditions were not favorable for diseases, and very little disease developed in plots. GLS was the most prominent disease and reached low severity. On both rating dates all Lucento treatment timings reduced GLS severity over the nontreated control, with the VT application having significantly less GLS compared to the R3 application on August 16 (Table 50). Harvest moisture was significantly lower in the nontreated control and V6 applications when compared to VT and R3 applications. No significant differences were detected between treatments for canopy greenness and yield of corn.

TABLE 50. *Effect of Fungicide Timing on Foliar Diseases Severity, Canopy Greenness, and Yield of Corn*

TREATMENT, RATE/ACRE, AND TIMING[z]	GLS %[y] AUGUST 2	GLS %[y] AUGUST 16	CANOPY GREEN[x] %	HARVEST MOISTURE %	YIELD[w] BU/ ACRE
Nontreated control	0.67 a	1.4 a	70.0	20.6 b	244.4
Lucento 4.17 SC 5.0 fl oz at V6	0.12 b	0.3 bc	71.3	20.5 b	248.9
Lucento 4.17 SC 5.0 fl oz at VT	0.01 b	0.1 c	74.2	21.5 a	257.1
Lucento 4.17 SC 5.0 fl oz at R3	0.22 b	0.7 b	72.1	21.4 a	257.7
P-value[v]	0.0029	0.0004	0.4606	0.0081	0.3450

[z] Fungicides were applied on June 15, July 19, and August 4 at the V6, tassel (VT), and milk (R3) growth stages, respectively. Treatments were applied at R3 contained a nonionic surfactant (Haf-Pynt) at a rate of 1.6 oz/Acre

[y] Disease ratings were assessed on August 2 at the milk (R3) growth stage and on August 16 at the dent (R5) growth stage. GLS = gray leaf spot

[x] Canopy greenness was visually assessed as a percentage (0–100%) of canopy green on September 2.

[w] Yields were adjusted to 15.5% moisture after harvest on October 5.

[v] All data were analyzed in SAS 9.4 (SAS Institute, Cary, NC). A generalized linear mixed model analysis of variance was performed using PROC GLIMMIX. Values are least squares means, and values with different letters are significantly different based on a least squares means test (α = 0.05).

FIELD-SCALE FUNGICIDE TIMING COMPARISON FOR FOLIAR DISEASES ON SOYBEAN IN SOUTHEASTERN INDIANA, 2022 (SOY22-08.SEPAC)

K. G. Waibel, J. R. Wahlman, A. Helms, and D. E. P. Telenko, Department of Botany and Plant Pathology, Purdue University West Lafayette, IN 47907-2054

SOYBEAN (*GLYCINE MAX* P38T05E)

Frogeye leaf spot, *Cercospora sojina*
Septoria brown spot, *Septoria glycines*
Downy mildew, *Peronospora manshurica*

A trial was established at the Southeast Purdue Agricultural Center (SEPAC) in Jennings County, Indiana. The experiment was a randomized complete block design with four replications. Plots were 30 feet wide and 700 feet long and consisted of 24 rows, and the two center rows were used for evaluation. The previous crop was corn. Standard practices for nonirrigated soybean production in Indiana were followed. Soybean cultivar P38T05E was planted in 15-inch row spacing at a rate of 130,000 seeds/acre on May 2. All fungicide applications were applied at 20 gal/acre and 35 psi. Fungicides were applied on June 15, July 19, and August 12 at V4, beginning pod (R3), and beginning seed (R5) growth stages, respectively. Disease ratings were assessed on August 16 at maturity (R6) growth stage. Frogeye leaf spot (FLS) was rated in the upper and lower canopies, downy mildew (DM) was rated in the upper canopy, and Septoria brown spot (SBS) was rated in the lower canopy. Severity of each disease was visually assessed as a percentage (0–100%) of symptomatic leaf area in the canopy in three locations in each plot on August 16. All ratings were averaged in each plot before analysis. Soybean plots were harvested on October 19, and yields were adjusted to 13% moisture. Data were subjected to a generalized linear mixed model analysis of variance performed using PROC GLIMMIX. Values are least squares means, and values with different letters are significantly different based on a least squares means test (α = 0.05).

In 2022 weather conditions were not favorable for disease development, and very little foliar disease developed in plots. FLS, downy DM, and SBS reached low severity. There were no significant differences between treatments for FLS in the upper canopy and SBS (Table 51). All treatments significantly reduced FLS severity in the lower canopy and DM severity compared to the nontreated control. No significant differences were observed for yield of soybean.

TABLE 51. *Effect of Fungicide Timing on Foliar Disease Severity and Yield of Soybean*

TREATMENT, RATE/ACRE, AND TIMING[z]	FLS[y] UPPER CANOPY %	FLS[y] LOWER CANOPY %	SBS[y] %	DM[y] %	YIELD[w] BU/ACRE
Nontreated control	0.5	0.4 a	0.6	0.2 a	68.1
Lucento 4.17 SC 5.0 fl oz at V4	0.1	0.1 b	0.3	0.4 b	68.4
Lucento 4.17 SC 5.0 fl oz at R3	0.0	0.0 b	0.6	0.7 b	71.9
Lucento 4.17 SC 5.0 fl oz at R5	0.0	0.0 b	0.7	0.3 b	72.7
P-value[v]	*0.0508*	*0.0448*	*0.2880*	*0.0093*	*0.1405*

[z] Fungicides were applied on June 15, July 19, and August 12 at V4, beginning pod (R3), and beginning seed (R5) growth stages, respectively, and contained a nonionic surfactant (Haf-Pynt) at a rate of 1.6 oz/acre.

[y] Foliar disease severity was visually rated on a scale of 0–100% of the upper and lower canopies with disease symptoms on August 16. FLS = frogeye leaf spot, SBS = Septoria brown spot, DM = downy mildew.

[x] Sudden death syndrome (SDS) was rated by visually assessing the percentage incidence in the canopy area at three locations in each plot on August 16.

[w] Yields were adjusted to 13% moisture after harvest on October 19.

[v] All data were analyzed in SAS 9.4 (SAS Institute, Cary, NC). A generalized linear mixed model analysis of variance was performed using PROC GLIMMIX. Values are least squares means, and values with different letters are significantly different based on a least squares means test (α = 0.05).

APPENDIX: WEATHER DATA

TABLE 52. Average Monthly Weather Conditions at the Purdue Agronomy Center for Research and Education (ACRE), the Pinney Purdue Agricultural Center (PPAC), the Southwest Purdue Agricultural Center (SWPAC), the Davis Purdue Agricultural Center (DPAC), the Northeast Purdue Agricultural Center (NEPAC), and the Southeast Purdue Agricultural Center (SEPAC) in Indiana, 2022[z]

MONTHS	ACRE			PPAC			SWPAC		
	TEMP. MIN.[y] °F	TEMP. MAX.[y] °F	TOTAL PRECIPIT.[x] (IN.)	TEMP. MIN.[y] °F	TEMP. MAX.[y] °F	TOTAL PRECIPIT.[x] (IN.)	TEMP. MIN.[y] °F	TEMP. MAX.[y] °F	TOTAL PRECIPIT.[x] (IN.)
January	12.5	32.4	0.47	9.8	27.5	0.25	19.3	37.3	2.30
February	18.3	37.5	2.28	16.9	33.8	1.72	24.1	44.3	4.30
March	32.8	52.8	3.40	29.2	48.3	2.86	38.2	58.4	4.36
April	39.6	59.8	2.74	35.1	54.6	3.09	44.8	64.3	4.98
May	55.5	75.8	5.77	51.5	71.9	2.72	58.7	78.5	4.87
June	60.0	84.4	1.20	57.2	80.9	2.11	64.9	87.7	1.39
July	64.0	84.5	1.74	61.2	81.4	3.58	69.7	87.6	13.18
August	61.0	83.3	4.47	58.7	80.4	3.55	66.2	86.3	2.39
September	53.1	77.5	1.80	51.3	74.2	1.34	58.6	81.3	1.16
October	40.4	65.7	2.73	38.2	62.7	4.09	44.6	69.4	1.29
November	33.4	52.8	1.97	30.8	49.7	1.18	36.8	56.5	1.45
December	24.5	38.4	1.25	20.9	34.2	1.06	27.6	42.6	3.02
Annual	41.4	62.2	29.82	38.5	58.4	27.55	46.3	66.3	44.69

MONTHS	DPAC			NEPAC			SEPAC		
	TEMP. MIN.[y] °F	TEMP. MAX.[y] °F	TOTAL PRECIPIT.[x] (IN.)	TEMP. MIN.[y] °F	TEMP. MAX.[y] °F	TOTAL PRECIPIT.[x] (IN.)	TEMP. MIN.[y] °F	TEMP. MAX.[y] °F	TOTAL PRECIPIT.[x] (IN.)
January	12.0	32.1	1.66	11.5	29.8	0.51	18.2	37.1	2.61
February	19.6	38.6	3.13	17.9	35.2	2.61	23.8	44.8	6.43
March	31.8	53.2	4.05	31.7	51.6	3.93	35.2	58.2	3.60
April	38.4	58.2	2.81	37.8	56.2	3.49	42.0	62.9	3.57
May	53.8	74.8	3.63	53.6	73.2	4.4	55.5	77.6	5.04
June	58.8	84.7	1.33	59.9	83.1	1.65	60.4	85.9	3.60
July	64.4	84.4	5.61	63.6	82.8	8.06	67.1	86.6	7.04
August	60.1	82.7	2.89	60.9	82.0	2.31	62.9	85.1	4.38
September	52.8	76.4	1.85	53.7	75.5	1.39	56.5	80.4	3.54
October	39.2	65.7	0.87	40.1	64.3	2.86	40.7	68.4	1.50
November	31.7	53.9	0.86	32.8	51.6	2.3	35.0	57.9	1.16
December	22.9	38.5	1.82	23.5	35.7	2.04	26.7	43.3	2.65
Annual	40.6	62.0	30.51	40.8	60.2	35.55	43.8	65.8	45.12

[z] Data courtesy of Indiana State Climate Office, Beth Hall, Jonathan Weaver and Austin Pearson, https://ag.purdue.edu/indiana-state-climate/. Taken from Purdue Mesonet stations.

[y] Average minimum and maximum temperatures for each month.

[x] Total precipitation for each month.

ABOUT THE AUTHORS

DARCY E. P. TELENKO is an associate professor and Extension plant pathologist in the Department of Botany and Plant Pathology at Purdue University. Her interdisciplinary research and Extension program are involved in studying the biology and management of soilborne and foliar pathogens of agronomic crops. Telenko is a native of western New York and received her PhD at North Carolina State University. She has published more than 60 peer-reviewed manuscripts and 200 Extension publications. She was awarded the 2024 Leadership Award from the Purdue University Cooperative Extension Specialist Association.

SUJOUNG SHIM is a research associate in the Department of Botany and Plant Pathology at Purdue University. Her research involves designing, conducting, analyzing, and reporting on a variety of research projects. She has a BS in pharmaceutical science and an MS in public health, both from Purdue University. Shim has served as a coauthor on more than 10 peer-reviewed publications and 25 peer-reviewed technical reports.

* 9 7 8 1 6 2 6 7 1 2 5 9 1 *